BIRD MIGRATION

BIRD MIGRATION

ROBERT BURTON

AURUM PRESS

First published 1992 by Aurum Press Limited
10 Museum Street, London WC1A 1JS

A catalogue record for this book is available from the British Library.

ISBN 1 85410 205 2

1993 1995 1996 1994 1992
1 3 5 7 9 10 8 6 4 2

An Eddison · Sadd Edition
Edited, designed and produced by
Eddison Sadd Editions Limited
St Chad's Court, 146B King's Cross Road
London WC1X 9DH

Phototypeset in Perpetua by Wyvern Typesetting Limited, Bristol
Origination by Scantrans, Singapore
Printed and bound by Dai Nippon, Hong Kong

FRONTISPIECE

*The knot is one of the world's great long-distance migrants. Here, a flock flying over an estuary
catches the sun as the birds twist and turn with marvellous co-ordination as a prelude to setting
off on the next stage of their journey.*

CONTENTS

PREFACE 6

INTRODUCTION 8

Migration observed 10 • What is migration? 11 • The mystery of migration 12
Where do birds go? 13 • Radio tags 17 • Built for travel 18 • Why migrate? 20

CHAPTER 1   TYPES OF MIGRATION   24

Migration patterns 26 • Problems of knots 28 • Partial migration 29
Differential migration 31 • European robins 32 • Loop migration 33
Leapfrog migration 36 • Irruptions 37 • Movements caused by weather 39

CHAPTER 2   PREPARATIONS FOR THE JOURNEY   40

Getting ready 42 • Juvenile explorations 43 • Fuel for the journey 44
The moult 46 • Moult migration 48 • Motive to move 49 • Time to go 50
Fair winds 52 • When to fly: night or day? 55

CHAPTER 3   FINDING THE WAY   58

Studying navigation 60 • Piloting, compass courses and navigation 62
Programmed routes 68 • Putting it all together 70 • Making a success of failure 73

CHAPTER 4   FLIGHT PLANS   74

Planning the journey 76 • The Great Circle route 77 • Difficult crossings 79
The greatest barrier 80 • Arctic terns: champion migrants 82 • Speed of flight 84
Flying high 89 • Economy travel 90 • Tracking migrating cranes 96
Journey time 98 • Staging posts 99 • Delaware Bay 101

CHAPTER 5   HAZARDS AND DELAYS   102

Salutary stories 104 • The scale of disaster 105 • Foul winds 106
Stray birds on the Scilly Isles 108 • Falls 110 • Going the wrong way 111
Predation pressure 112 • Human threats 114

CHAPTER 6   WINTER HOMES   116

Winter life 118 • Packing them in 121 • How far to go 125 • Winter waders 126
Movements in the tropics 128 • American wood warblers 134

CHAPTER 7   THE RETURN   136

The race to get back 138 • On their marks 140 • Geese to the Arctic 142
Early returns 144 • Returning sequence 147

LIST OF SCIENTIFIC NAMES   154
FURTHER READING   156
INDEX   157
ACKNOWLEDGEMENTS   160

PREFACE

The British naturalist and author W.H. Hudson told the poignant story of a pair of upland geese in Argentina, where he spent his early years. Some days after the flocks had flown south at the end of winter, a pair was spotted on the plain. The male was alternately walking and flying in a southerly direction, but repeatedly turning back and calling to his mate who was crippled with a broken wing. So great was the urge to migrate to the breeding grounds that she was attempting the journey on foot.

Since earliest times the migration of birds has had a powerful impact on human consciousness. The annual cycle of arrivals and departures was important for people whose lives were intimately geared to the changing seasons, and even in modern times the appearance of favourite birds at the end of winter brings pleasure to lives that are insulated from the caprices of nature. Bird migration is also the most striking and easily observed example of the imperative of instinct driving the lives of animals. It has always been a source of wonder how birds travel to the ends of the world and back, navigating efficiently by day or night and coping with long sea crossings and adverse weather.

Because we are most aware of migrant birds when they arrive in spring and depart in autumn, and may have little thought about what happens during the intervening months, our view of migration can easily become too simplistic – the birds are merely obeying blind instinct. But the truth is much more complex and much more fascinating, and the explanations, as they are uncovered by often ingenious scientific research, so much more satisfying.

It is impossible to describe the whole complexity of bird migration. Even to attempt a description of the routes that birds take would result in a diagram as intricate as that of a computer's circuitry. To reduce such a bewildering mass of information, I have written this book from the viewpoint of readers in Europe and North America and describe the movements of birds from the north temperate zone to the tropics and beyond. I have ignored migrations through Asia and the Pacific which are still relatively poorly understood. I have also given the impression that all birds head south after breeding but some, like the upland geese described above and the double-banded plover that winters in Australia and nests in New Zealand, travel in the opposite direction.

New facts about bird migration are continually being discovered. In 1985, large numbers of knots were found to use Balsfjord, northern Norway, as a staging post in their flight from western Europe back to their breeding grounds in Greenland and north-east Canada.

INTRODUCTION

UNRAVELLING THE MYSTERY

The seasonal ebb and flow across the face of the earth that is bird migration fires the imagination. We wonder at these frail scraps of life battling their way across ocean and desert, against all odds, and we marvel at the skills that enable them to find their destinations. And the scale of migration is enormous. All over the world, hundreds of millions of birds pour across land and sea in a twice-yearly urge to seek new homes.

The turnstone is one of the many species of waders that undertake amazingly long journeys twice every year.

MIGRATION OBSERVED

Much of the mass movement of bird migration is a ghostly, unseen passage. In spring the woods are suddenly full of singing warblers that were not there a few days previously. Hordes have crept in unnoticed. We have missed them because they move by night and travel at heights that render them invisible. Sometimes we can watch the passage of day-flying hawks overhead or seabirds streaming around a headland and, counting the numbers per hour, get some idea of the scale of migration.

The arrival of flocks of birds, such as these bar-tailed godwits, is welcomed by bird-watchers as a sure sign that migration is under way. The movements of the flocks provide one of the great spectacles of natural history. As they flood in, there will be many birds to be identified, including perhaps some surprise rarities. Their departure is less noticeable, as the numbers of birds gradually dwindle.

The sight and sound of these travellers is immensely evocative, as the poet Tennyson crystallized in the lines:

And fainter onward, like wild birds that change
Their season in the night, and wail their way
From cloud to cloud, down the long wind
Shrill'd; but in going mingled with dim cries
Far in the moonlit haze among the hills.
THE PASSING OF ARTHUR, 1842

W.H. Hudson told of the distinguished clergyman who retired, at the turn of the century, to a lonely village on the most desolate part of the east coast of England. His explanation was that it was the only spot where, sitting in his front room, he could listen to the cry of pink-footed geese arriving for the winter. 'Only those who have lost their souls will fail to understand' was Hudson's comment.

WHAT IS MIGRATION?

Bird migration is usually taken to mean the seasonal move-ment of birds at predictable times of the year between breeding and non-breeding areas, although it is sometimes extended to include any significant movement by a bird from its usual place of residence. The first, more specific, definition is the one adopted in this book.

In the tropics, uniform conditions and an abundance of food all year enable some species to stay put. Elsewhere, a very high proportion of the world's species of birds migrate, but populations of some species may remain in one country all year. Even the common birds that appear to be always with us are subject to the ebb and flow of migration. Because some individuals remain in one place the arrival and departure of their relatives easily go unnoticed. The house sparrow almost qualifies as a complete stay-at-home but there is a migratory race in central Asia. Many species of owls, woodpeckers and gamebirds are sedentary but there are some surprises among other bird groups. The familiar and widespread wren, known in North America as the winter wren, that is usually seen flying from bush to bush on small, whirring wings, may migrate considerable distances: European wrens cross the North Sea and Canadian wrens fly to the Gulf of Mexico. The corncrake that skulks in dense herbage and rarely takes wing, even when disturbed, migrates from north-west Europe to Africa while even the tiny rufous hummingbird, which weighs hardly more than 3 grams ($\frac{1}{10}$ ounce), flies from Alaska to southern Mexico.

The common nighthawk, a member of the nightjar family, is more often heard than seen. It feeds on night-flying insects and leaves North America for the tropics when the insects disappear in autumn.

Birds are always on the move, but the term migration is usually restricted to annual journeys. A migration need not involve a journey across countries or continents. The North American blue grouse, for instance, nests in deciduous woods and migrates 300 metres (1,000 feet) up into mountain pine forests for the winter. Prairie falcons migrate east–west, from the Rocky Mountains to the prairies. They return to the mountains when their ground squirrel prey emerges from hibernation and leave in June or July when the squirrels retreat below ground again to escape the midsummer heat. Also, a migration need not involve the whole of a species' population. Even a few individuals of that archetypal migrant, the swallow, stay behind for the winter in temperate countries.

We think of migration as being southwards to the sun in the autumn and back to nest as the climate warms in the spring, but this belies the bewildering complexity of move-ments, as well as of techniques and strategies evolved by the birds to ensure that they arrive safe and sound. However, in this book we will cut through the interwoven, overlapping web of bird migrations and concentrate on the major, and best-studied, movements between Europe and Africa and between North America and Central or South America.

THE MYSTERY OF MIGRATION

The 'mystery of migration' is a common phrase. It has a nice, alliterative rhythm but there is nothing unique about migration. Like all other phenomena in the natural world, it is a progressive mystery whose solution will never be final. We know much more about where, why and how birds migrate but, as knowledge accumulates, each question that is answered poses new ones. It is possible to write a short description of bird migration that will read as if the subject is completely understood, but this is true only in generalities. Only when there is space for details can it be appreciated that underneath this apparent certainty, there is a labyrinth of unresolved problems for researchers to tackle.

There are four major topics in understanding the 'mystery of migration'. The first is the question of the birds' destination. In the many centuries since people became aware of the arrival and departure of birds with the changing seasons, they must have wondered where they came from and where they went to. When the extent of migration became known, their remarkable ability to navigate with the precision of the best human navigational instruments was revealed. There is also the problem of how birds, even the smallest species, physically manage to undertake journeys covering thousands of kilometres as a matter of routine. Finally, there is the basic and most profound problem of *why* birds migrate.

Scientific research and observations have explained much of what was baffling. Over the last 50 years we have gained a better idea of how and why birds migrate. We can see the advantages of what appear to be almost suicidal flights that take up a large part of a bird's life, and we can also see that the flights are not such hopeless ventures as they might seem.

Yet much of the detail is still unexplained and any discussion of bird migration has to be hedged with question marks. This is partly due to the difficulties of studying birds on their journeys. It would be very useful, for instance, to have an accurate figure for the fuel consumption of a small bird on a non-stop sea crossing. The obvious way to find out would be to weigh it just before it takes off and again after it lands but, except in rare instances, it is impossible to tell if a bird is just about to depart or, at the other end of the journey, if it is newly arrived – let alone find an individual bird at the end of its flight. The best that can be done is to weigh a sample of birds as near as possible to departure time and again as soon as possible after they have arrived. This will give a good average for the fuel consumption.

Much can be learned about bird migration by watching the movements of birds. The mass movements of seabirds, for instance, can be observed as the birds are channelled past headlands.

WHERE DO BIRDS GO?

In some cases they migrate from near at hand; in others they may be said to come from the ends of the world, as in the case of the crane, for these birds migrate from the steppes of Scythia to the marshlands south of Egypt, where the Nile has its source.
ARISTOTLE, THIRD CENTURY BC

The first great question in the 'mystery of migration' is: where do the birds go? Aristotle knew the destination of the crane but many birds disappeared to places beyond the then known world. In north-west Europe the nesting place of barnacle geese was unknown and it was believed that the flocks that arrived each winter had developed from goose barnacles, the crustaceans that suspend themselves from ships' timbers and driftwood. In 1584, the English scholar William Harrison described picking barnacles off the hulls of ships newly arrived in the River Thames and opening them to find 'the proportion of a fowl in one of them more perfectly than in all the rest, saving that the head was not yet formed'. The truth of the origin of barnacle geese became known in 1596 when a Dutch expedition, under the command of Willem

People once had bizarre ideas about where barnacle geese came from. It is now known that they arrive on the coasts of Europe in the autumn from their Arctic breeding grounds.

Barents, discovered Svalbard and found colonies of nesting barnacle geese.

Unfortunately, Aristotle also introduced two red herrings – hibernation and transmutation – which led naturalists astray for two millennia. Swallows, kites and turtle doves were believed to hibernate through the winter, and redstarts (summer visitors to Aristotle's Greece) to become transformed, or 'transmuted', into robins (winter visitors) in autumn.

— DIRECT OBSERVATION —

It is not surprising that such strange notions of hibernation and transmutation should have been held by intelligent people, especially since many birds migrate by night so that no one noticed their appearance and disappearance. The migrations of small birds posed a problem which was eventually resolved mainly by two methods of study. One was systematic bird-watching, which still provides huge amounts of information as observers gather at favourite sites to watch passing birds. These places are usually bottlenecks where migrants are funnelled around mountains and headlands, or along river valleys and shorelines. Thousands may stream past in a few hours, in an unforgettable 'rush'. Small islands attract large numbers of migrants to rest on their sea crossings, sometimes arriving in a 'fall' after adverse weather (see page 110).

The pioneers of migration study started systematic observations at these special locations, which sometimes became the sites of established bird observatories. The first and most famous was that set up by Heinrich Gätke on the island of Heligoland on the east coast of the North Sea. Also famous is the Scottish island of Fair Isle, which is a haven for birds crossing the North Sea. In North America, Cape May, Connecticut, is famous as a place for watching departing migrants in the autumn, while Scandinavian birds are funnelled through Falsterbo in southern Sweden for their passage across the Baltic. Not all observatories are coastal: the Col de Bretolet on the Swiss/French border is a main thoroughfare for migrants passing through the Alps, and in the USA Hawk Mountain, Pennsylvania, attracts large crowds to watch the impressive passage of birds-of-prey.

MOON-WATCHING AND RADAR

Direct observation has its shortcomings. Each observatory is only a tiny 'window' overlooking the swirling panorama of bird migration. Most of the birds will pass overhead out of sight and hearing, but the view can be improved by counting the silhouettes of birds passing the disc of the moon or the beam of a searchlight, a technique developed by the American George H. Lowery. In October 1952, more than 1,000 observers across North America spent four consecutive nights 'moon-watching', and they revealed a broad stream of birds

Attracting birds to a hide not only gives bird-watchers a superb view but also enables researchers to keep a record of their numbers and even to recognize individuals and keep a history of their movements.

heading southward across the eastern half of the continent in the northerly wind that followed behind a cold front.

Anyone can try 'moon-watching', but serious studies of migratory movements now use radar. When radar was first used to detect enemy aircraft approaching the shores of Britain in the Second World War, the operators were confused by 'angels' – echoes coming from areas where there were no aircraft. Among the causes were flocks of birds. Radar proved to be an excellent means of plotting the movements of migrants, which appear on the screen either as images travelling across an outline of the landscape or as displays giving range, height and numbers of birds. Some sensitive radars even make it possible to identify individual species by the patterns of their wingbeats. Radar is still limited in its range, and interpretation of the observations has often proved difficult, but this method gives good records of mass movements.

BIRD-RINGING

The migratory habits of individuals over long distances and wide spans of time are studied by bird-ringing, known in North America as bird-banding. Threads tied around birds' legs, or metal rings placed around their necks or legs, have at

The Heligoland trap is a permanent structure used at bird observatories to trap birds for ringing. The birds are lured to the entrance by the offer of food or shelter and then carefully driven into the funnel. A swing door is closed to prevent them flying out again and eventually they take refuge in the box at the end. The bird-ringer reaches into the box and removes the birds one by one.

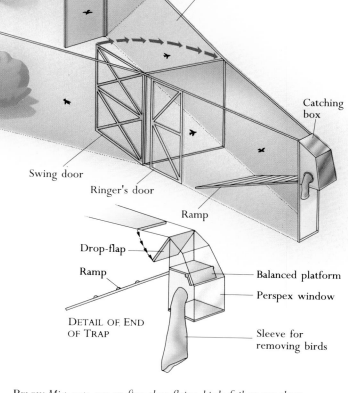

Funnel

Catching box

Guide wall

Baffle

Swing door

Ringer's door

Ramp

Drop-flap

Ramp

DETAIL OF END OF TRAP

Balanced platform

Perspex window

Sleeve for removing birds

various times over the centuries conveyed messages or demonstrated ownership of birds. The earliest reference to the study of bird migration is the thirteenth century account, no doubt fictitious, of a swallow which had a parchment tied to its leg, bearing the message: 'O Swallow, where do you live in winter?' Next spring it returned with the answer: 'In Asia, in the home of Petrus'.

More reliably, John James Audubon, the famous early nineteenth-century American pioneer ornithologist and artist, tied silver threads around the legs of eastern phoebe nestlings and caught them next year as nesting adults. Several other experiments were tried but systematic bird-ringing was started by the Danish schoolmaster Hans Christian Mortensen, in 1899, when he fitted 164 starlings with aluminium bands or rings, each with a serial number and a return address. This individual number that identifies a bird for life was the most important advance. Other schemes followed and eventually coalesced into national programmes. North America has a single bird-banding scheme run by the US Fish and Wildlife Service, while EURING, the European Union for Bird Ringing, standardizes and co-ordinates national schemes on the other side of the Atlantic.

The principle of bird-ringing is that each ring bears a unique reference number and the address of the scheme that issued it. The bird-ringer records the species of bird, the place where it was ringed and data on the condition of the bird: its sex and age (if known), the state of its plumage, its weight and various body measurements. This alone gives researchers a vast amount of information. Some time later, although the chances are usually remote, someone may find the ringed bird again. It may be seen by a bird-watcher, shot by a hunter, caught in a fishing net or brought in by the cat.

Hopefully, the ring will be seen, removed and returned, along with information on date, place and how the bird was found. If the bird is caught by a ringer, the number on the ring is carefully read, and the weight, measurements and other

BELOW *Mist nets are so fine that flying birds fail to see them against a dark background. They hit the net and drop into the loose folds where they become entangled. Extracting the birds so that they can be ringed is a skilled operation.*

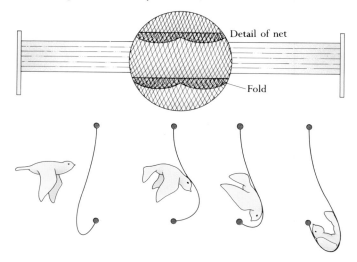

Detail of net

Fold

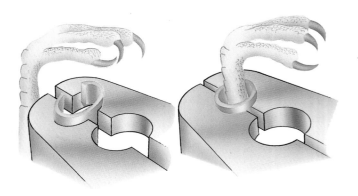

LEFT A combination of coloured plastic rings allows researchers to identify a bird and study its movements and behaviour without capturing it more than once.

ABOVE A ring, made of tough but malleable aluminium alloy, is closed with a pair of special pliers, then rotated and tightened with a final squeeze. It is so light in relation to the bird's weight that it is no more of a burden than a wrist-watch is to a human.

records are repeated and this additional information also returned. The rings are so small that only a shortened address can be accommodated. British rings read 'Inform Brit Nat Hist Mus London' and North American bands are inscribed 'Fish and Wildlife Service Washington DC USA'. The addresses are abbreviated even more on the smallest rings. Russian rings are labelled 'MOSHWA', a word that does not exist but which is intelligible as the capital of Russia to people familiar with either Cyrillic or Roman letters.

Bird-ringing has revolutionized our knowledge of migration by revealing the destinations of migrants. The information sometimes comes to light in curious ways. Imagine the excitement of the ornithologist studying the feeding habits of the gyrfalcon on Ellesmere Island, in the far north of Arctic Canada, who teased apart a pellet containing the inedible remains of its prey and found a bird-ring. It had been attached, five years earlier, to the leg of a knot wintering in England. A stranger record came from a Norwegian who, in 1970, tamed an oystercatcher so that it would come at his call and perch on his hand to be fed. One day in 1976, he realized it had a ring on its leg. Since the bird was so tame, the man could read the number and address. His oystercatcher had been to England and back.

Unless a ringed bird can be lured very close to an observer with binoculars or telescope, it must be caught for the number to be read. This drawback can be overcome by using coloured or numbered plastic rings, or by attaching plastic tags to the wings which can be read with binoculars at a distance. If only general information is required about the destination of a bird in a particular population, they can be marked with dye and bird-watchers at likely places asked to keep watch for any luridly coloured birds. A disadvantage of this method is that

the dye disappears as the bird moults it feathers.

Sometimes information about movements can be gained from close examination of an unmarked bird. Bird-ringers catching dunlins around the coasts of western Europe in winter can distinguish birds that nested in north-east Greenland or northern Scandinavia from those coming from breeding grounds in north-west Russia by subtleties of plumage and body measurements.

One reason for the success of bird-ringing is that it is an absorbing hobby for large numbers of people. The time spent by bird-watchers licensed to catch, ring and record birds provides ornithologists with invaluable data for their studies. In the British Isles 800,000 birds are ringed each year, while in North America the total is over 1.5 million. A strict system of apprenticeship and licensing ensures that no harm is done to the birds.

Apart from the pleasure of handling and examining birds, especially rare species, as a 'ringer' you can experience the excitement of catching a bird ringed in a foreign country or of receiving a note that one of 'your' birds has been found in a distant part of the world. Some birds are found repeatedly, so giving a better picture of their migrations, but ringing reveals only isolated points in a lifetime of travels. The ultimate aim is to trace the exact movements of individual birds. The most exciting way of doing this is to follow them in light aircraft. This was tried by Harald Penrose, a test pilot, who intercepted swallows as they flew along the south coast of England, but the birds were too small and slow to track. Colin Pennycuick and Thomas Alerstam had better luck with migrant cranes, as will be described on pages 96–97. An even more promising technique is to attach to the bird a miniature radio beacon that continuously broadcasts its position.

RADIO TAGS

The revolution in the electronics industry has given ornithologists a marvellous new tool, the radio tag. This is a tiny transmitter that acts as a beacon so the bird can be tracked continuously with great precision. The first studies were made by following the tagged bird by car or plane. Keeping track of a transmitter with a limited broadcasting range was often hectic, but radio tags are now powerful enough to be tracked by satellite and their position fed to researchers in their offices. The data from radio-tagging is beginning to provide exciting information, and help to answer questions that had been very difficult to fathom by other means. At last, it is now possible to get a complete account of the travels of individual birds.

For instance, radio-tagged Canada geese have been tracked throughout their migration between their nesting grounds in Manitoba and their wintering grounds in Minnesota. The route along the border between forest and prairie was staked out by observers and the movements of the geese followed closely to give a remarkable picture of their migration.

The geese often flew the 855 kilometres (531 miles) non-stop in a flight lasting less than 12 hours. They chose to start in clear skies and with tail winds, but sometimes had to stop if they encountered snow or rain, or were caught by nightfall. While airborne, the geese made a bee-line for their destination at airspeeds averaging 60–65 kilometres (37–40 miles) per hour, and sometimes as much as 100 kilometres (62 miles) per hour. Plotting the birds' tracks over the ground showed that the geese were able to adjust for side winds by altering their heading to prevent drifting off-course. They probably do this by keeping watch on landmarks, which may explain why they prefer flying in good weather.

Although expensive, miniature radio transmitters, like the one fitted to this peregrine, give unrivalled scope for following the movements of birds. With the aid of satellite links, birds can be followed across oceans.

BUILT FOR TRAVEL

A cartoon by the American Gary Larson shows a flock of geese flying in V-formation. Below them is another formation, walking in the same direction. 'Say', says its leader, looking up at the flying geese, 'Look what *they're* doing'. The idea of birds migrating on foot is, of course, to be laughed at (except for Adélie and emperor penguins waddling over the frozen Antarctic Ocean to their breeding colonies) but the cartoon makes an interesting point. The key to avian migration is the power of flight, which allows birds to travel long distances economically. At first sight, this is puzzling. Flying is the most strenuous form of locomotion in the animal world. A bird in steady flapping flight uses 10 to 15 times as much energy as when perching. However, the important point is the transport cost. Less energy is used to move one gram of animal over a distance of 100 kilometres (62 miles) through the air than on the ground. Put another way, it is easy to imagine a swallow, which weighs only a few grams, travelling 100 kilometres (62 miles) but not a mouse, which weighs roughly the same as the swallow. The reason is that although flying is strenuous, the journey is very quick.

The best fuel to supply the energy for flight is fat. It releases more energy than carbohydrate or protein – 1 gram ($\frac{1}{28}$ ounce) of fat yields 9 kilocalories of energy compared with 4 kilo-

calories for carbohydrate – and it is lighter because it can be stored without the addition of water. Carbohydrate is stored in the form of the substance called glycogen, which has to be chemically bonded with three times its weight of water. The net result is that a store of fat contains eight times as much energy as the same weight of carbohydrate. Indeed, weight for weight, fat is a better fuel than petrol. Another advantage is that fat can be laid down in large masses under the skin and around the digestive organs. (Blowing on the breast of a small migrating bird to part its feathers may expose the yellow appearance of the fat under its skin.) So whatever its diet, from insects or fish to berries or seeds, the migrant processes its food into easily stored fat.

The vital factor for the migrant bird is the distance it can fly on its store of fat. This is one of the most amazing facets of bird migration. As much of a key to the 'mystery of migration' as the ability of a small bird to navigate across the world with pinpoint accuracy is its capacity to fly, non-stop, for a distance of 3,000 or more kilometres (1,860 or more miles) and remain

Despite their great size, Bewick's swans fly far and fast on migration. Their flight is fuelled by the large amounts of fat which they have stored in their bodies in preparation for the journey.

airborne for three or four days. These figures are based on calculations from the theory of bird flight aerodynamics, but there is plenty of evidence that birds do achieve flights of these ranges and durations in real life.

The secret of this apparent miracle lies not so much in the amount of fat that a bird can store as in the fact that flight range is dependent on the *proportion* of fat to body weight, and not upon the *absolute* fuel load. In other words, when a sedge warbler puts on 10 grams ($\frac{1}{3}$ ounce) of fat at Lake Nakuru, Kenya, in spring, it can fly northwards non-stop to the Mediterranean coast of Egypt because the fat represents 50 per cent of its total body weight. Ten grams of fat is only 0.1 per cent of the weight of a whooper swan, however, and would not propel it very far!

When the importance of the relationship of fuel load to body weight is realized, long-distance migration ceases to be a miracle. Doubling the lean body weight, that is having 50 per cent as fat, before leaving on migration is not unusual in small birds. They have an advantage over large birds because they have the power to carry the extra fat. Doubling the weight increases the power needed to fly by a factor of three. Large birds such as geese and swans do not have the spare power needed to lift such a heavy load of fuel into the air. However, their advantage is that they fly faster, accomplishing the journey quicker, so they spend less energy on life-support during the shorter flight. The result is that the record-breaking non-stop migrants are medium-sized birds that combine large fuel capacity with fast flight, such as the waders which fly from the Arctic tundra to the tropics and far beyond.

Carrying a load of extra fuel has two drawbacks. More power is needed to keep a heavy bird airborne and also to overcome the increased drag caused by its plump outline. This can be partly offset by strengthening the flight muscles, and some songbirds that cross the Sahara increase the weight of

Despite its small size, a sedge warbler can fly hundreds of kilometres non-stop because it can carry half its body weight as fat. Its capacity for long-distance flight is aided by an efficient physiological release of energy from the fat.

their flight muscles by 20 per cent. The extra muscle may act as an emergency fuel. If the bird has been caught by head winds over the sea and has to keep airborne, it can use muscle protein to keep on flying after all its fat has been exhausted. This is not such a desperate measure as it may seem because, with the heavy fat load gone, flying requires less power and the bird does not need so much muscle. The flight muscles are also adapted for the extra effort of long-distance flight by becoming more efficient at metabolizing fat, in much the same way as athletes' muscles during endurance training.

Fuel for flight is burnt, or oxidized, in the flight muscles and requires a steady supply of oxygen and the removal of carbon dioxide. Birds have a unique breathing system, in which the lungs are linked to three pairs of thin-walled air-sacs. These cause air to pass through the lungs in a single direction instead of in and out, as in the tidal, 'bellows' system of mammals. The one-way flow of air ensures more efficient extraction of oxygen and flushing of carbon dioxide. However, some bats, with typical mammalian lungs, are long-distance migrants, so the birds' airsac system would appear to be unnecessarily complicated. The advantage appears to be that it allows birds to fly at great heights. Mice in a test chamber were barely able to crawl at a pressure equivalent to 6,100 metres (20,000 feet), while sparrows could still fly easily.

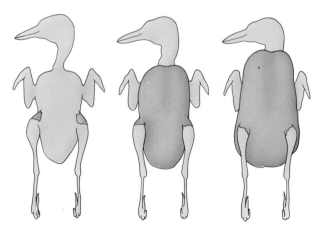

ABOVE *The amount of fat that a small bird can carry varies enormously. After a few days without eating it is extremely lean and will soon die if it cannot feed. With plentiful food it very rapidly puts on weight, laying down fat (shown here in red) that will power it for a long flight.*

WHY MIGRATE?

Some creatures can make provision against change (of season) without stirring from their ordinary haunts; others migrate, quitting Pontus and the cold countries after the autumnal equinox to avoid the approaching winter, and after the spring equinox migrating from warm lands to cool lands to avoid the coming heat.

ARISTOTLE, THIRD CENTURY BC

Although, as so often, Aristotle, the great Greek philosopher and 'Father of Natural History', had the first word on the subject, he was mistaken in believing that the purpose of migration is to avoid excesses of cold and heat. Birds can cope with extremes of temperature and the northern grouse called ptarmigan remain on the Arctic islands of Svalbard through the winter, experiencing temperatures as low as –30 degrees Centigrade (–22 degrees Fahrenheit), because they can find enough food in the form of plant shoots to survive.

The key to migration is food but, as food supplies are usually linked to climate, Aristotle can be forgiven for seeing a direct connection between the movements of birds and the changing seasons. In temperate and Arctic regions flying insects largely disappear in the winter. So aerial insect-eaters, such as swallows, swifts, flycatchers and nightjars, depart for warmer places. They are accompanied by warblers which pluck insects from leaves, but the tits and treecreepers that pick hibernating insects from crevices remain behind. Other insect-eaters turn to a diet of seeds or fruit. For instance, the tits and woodpeckers supplement their diet with seeds and the bearded tit switches to a diet of reed seeds, undergoing a major change to its alimentary system to cope with this less easily digested food.

Nevertheless, to say that birds migrate to avoid a seasonal shortage of food betrays a view of the bird world that is seen from a northern, European or North American, perspective in which birds migrate away to the tropics from the harshness of winter. The alternative is the southern viewpoint in which the question of why birds migrate is stood on its head. If they have space and resources on their wintering ground, why fly north in spring to breed elsewhere?

From either perspective, the key factor is the availability of resources. The northern summer gives an abundance of food, which can be gathered with less competition from the other species, and safe nesting places where there are fewer predators. After the summer, the food supply fails. This is most obvious in latitudes where food is hidden under ice and snow and most birds are forced to migrate. They withdraw to milder climates where there are enough resources for survival. The history of the serin provides a good illustration. This finch

Great tits are generally sedentary but large numbers leave northern Europe in years when a population build-up coincides with a failure in the crop of beechmast, the birds' main winter food.

ABOVE *The whinchat escapes European winters by flying to Africa but the journey time reduces the length of its breeding season.*

LEFT *Although capable of surviving cold winters, black-capped chickadees from northern parts of the species' range fly south.*

was confined to the Mediterranean region until about a century ago. Then it began to spread northward through Europe and it has now reached southern Sweden. In its ancestral home the serin is resident the year round, but in the far north of its new range it is migratory. It is forced to fly south because the weed seeds that it eats are covered with snow.

The southern perspective is that migration allows birds living in the tropics or temperate regions to take advantage of a seasonal abundance in the northern summer. This is most easily appreciated when we consider those birds that winter in temperate countries and migrate to the Arctic tundra to nest in a very short summer season of abundance. If birds did not migrate northwards, huge areas of land and untold resources would remain unexploited. There is generally competition among birds for food, so a move north into relatively empty country will give a bird an advantage over its sedentary relatives. It is not surprising, then, that there has been a pressure for the evolution of the migratory habit. The dickcissel is a bunting that winters on the grasslands of northern South America, where it competes with other birds for grass seeds. Migration to the North American prairies can be explained as an escape to an easier life where there are fewer seed-eaters to dispute for the resources.

Migration is not an easy undertaking and there are many risks and penalties. Whether or not a migratory life-style is worthwhile depends on the benefits it confers in either the summer or winter home outweighing the costs and dangers of the flight. The advantage of migration to a swallow is obvious – there are too few flying insects to support it through the winter on its breeding grounds – but, for many birds, the decision to migrate requires a resolution of the relative merits of staying and moving.

The advantages of migration are more far-reaching than the simple survival of the individual. They are important for the outcome of its nesting attempts. Take two very similar European species, the stonechat and whinchat. They are small members of the thrush family with very similar habits. Both live in grasslands and heaths and have a diet of insects. Yet the British population of stonechats is resident, or migrates no further than south-west Europe, while the whinchat migrates across the Sahara into tropical Africa.

Cold winters cause a high mortality among stonechats but leave whinchats unaffected. The advantage would seem to be with the whinchats but the stonechats quickly recoup their losses after a severe winter. By staying on or near the breeding ground they start nesting earlier in the year, and lay three clutches to the whinchats' two. Stonechats are also more successful at nesting and produce twice as many young. Both species thrive; the migratory whinchat fares better in cold years but the sedentary stonechat eventually catches up. It cannot be said that migration is a fundamentally 'better' strategy. It all depends on circumstances.

CHAPTER 1
TYPES OF MIGRATION

The essential function of migration is that it allows birds to exploit two, or more, habitats each year. When life becomes difficult in one home, birds fly to an easier life in another. Migration is so much a part of the life-style of birds (few species never migrate) because their power of flight lets them travel efficiently over land and sea. This chapter looks at the variety of different migration patterns that have been identified in birds.

Part of a huge flock of knots on its wintering ground. This wader is a champion long-distance migrant.

MIGRATION PATTERNS

One of the marvels of migration is the distance travelled by some migrants. Many go only as far as necessary to avoid the privations of winter. For example, a variety of birds – woodpigeons, wrens, robins, finches, crows and others – leave Scandinavia for milder winters in southerly parts of Europe, while in North America there is an exodus from the pine forest belt into New England and warmer regions further south. On the other hand, some birds travel much further and fly to the tropics and beyond. Among the waders, for instance, there are those that spend the winter in temperate countries, either because they are remaining near their nesting places or because they have migrated down from the Arctic. Others, even of the same species, fly on to wintering grounds on the far side of the equator.

One advantage of flying to warmer countries to escape the winter is that less energy is needed to keep warm. Even small daily savings of energy in the warmer habitat mount through the winter months to offset the cost of two long flights. Yet many of these flights seem to be absurdly long. When looking at maps of migration routes, such as that of the knot (see page 28), the obvious question is why do some birds fly so far? Surely they could escape the hardships of Arctic or even temperate winters with a much shorter flight to the milder regions of southern Europe or North America, instead of

ABOVE *One of the great puzzles of migration is why some birds, such as these bar-tailed godwits, migrate so far. They may not be able to accomplish their journeys without help from tail winds.*

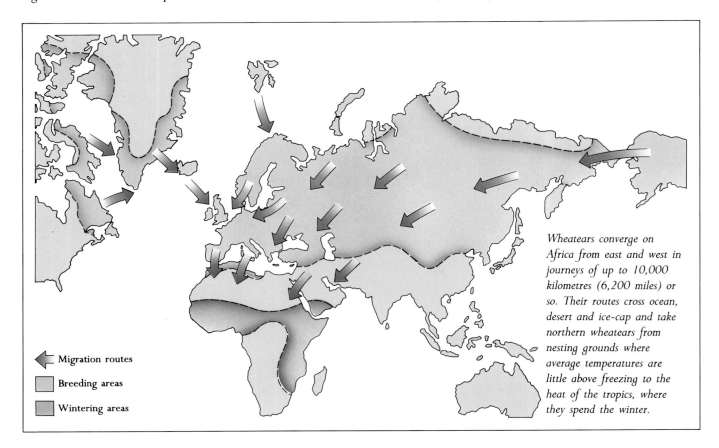

Migration routes

Breeding areas

Wintering areas

Wheatears converge on Africa from east and west in journeys of up to 10,000 kilometres (6,200 miles) or so. Their routes cross ocean, desert and ice-cap and take northern wheatears from nesting grounds where average temperatures are little above freezing to the heat of the tropics, where they spend the winter.

ABOVE *This family of whooper swans, two adults and two cygnets (foreground), stayed together when they migrated from the Arctic. The young will benefit from their parents' experience of the route and will be protected until capable of fending for themselves.*

crossing the equator? Some birds, indeed, do just this. The advantage of long-distance migration is a subject that has received very little study and there is no simple answer. It seems odd, for example, that Finnish buzzards migrate all the way to South Africa, yet their neighbours in Sweden fly only as far as western Europe.

The exact routes taken by migratory birds are almost impossible to illustrate. There are at least as many routes as there are bird species, and within a single species there are variations in timing and pathways. When there have been sufficient ringing returns to plot the paths taken by one species, the result is a cartographer's nightmare of tracks interweaving and overlapping in space and time.

In real life there is no neat division into summer breeding and winter non-breeding ranges linked by an arterial route that can be easily drawn on a map. One fraction of the population may migrate farther than another, or not at all, and it may follow its own unique itinerary and travel at a different time. Any attempt to describe a bird's migration must of necessity be a simplification. An arrow on a map is only a general indication of a route. Most species migrate on broad fronts that may extend hundreds of kilometres, and are concentrated into a stream only when they are funnelled around mountains, along coasts or across straits. For instance, hawk migration is normally spread over eastern North America in the autumn but west and north winds drift the birds laterally until they meet mountain ridges of the Atlantic coast. The result is impressive numbers of hawks passing Hawk Mountain, Pennsylvania, and Cape May, New Jersey. Watching them at such places can be very exciting but gives a false impression of the overall pattern of hawk migration.

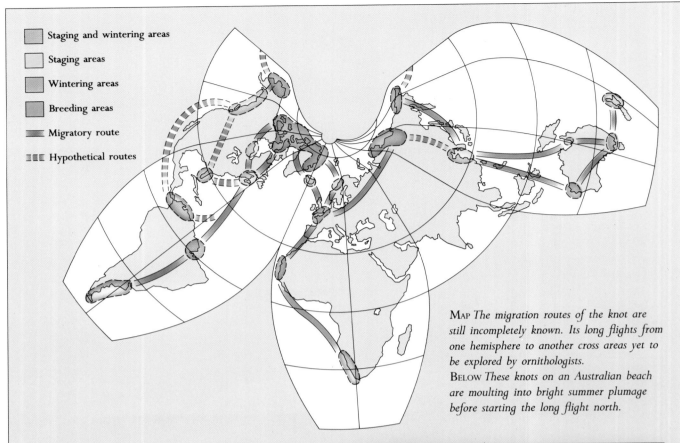

Staging and wintering areas

Staging areas

Wintering areas

Breeding areas

Migratory route

Hypothetical routes

MAP *The migration routes of the knot are still incompletely known. Its long flights from one hemisphere to another cross areas yet to be explored by ornithologists.*
BELOW *These knots on an Australian beach are moulting into bright summer plumage before starting the long flight north.*

PROBLEMS OF KNOTS

The plump little wader called the knot, or red knot in North America, leads a double life. It nests in the short Arctic summer on the bleak tundra, where its widely spaced nests and elusive habits are frustrating to ornithologists trying to study it. Before winter returns, it migrates south to estuaries and coasts where it lives in large and very conspicuous flocks, which perform spectacular aerial manoeuvres. Over several decades, thousands of these wintering birds have been caught and ringed, and many of these have been recovered on their distant breeding grounds, back at the wintering sites and at staging posts in between. The result is that ornithologists have been able to build up a picture of the movements of one of the world's major migrants, a bird that is capable of 3,000-kilometre (1,850-mile) non-stop flights.

The map on this page shows the enormous extent of the knot's migrations: from the most northerly tundras bordering the Arctic Ocean to the tips of continents stretching towards Antarctic regions. Considering the remoteness of some of the breeding grounds, wintering areas and intermediate staging points, it is hardly remarkable that there are some parts of the migration picture that are still indistinct or remain to be drawn in when more information is available. The routes taken by knots through North America, for example, remain hypothetical. Until 1985 it was believed that the entire Greenland/north-east Canada population flew back from the North Sea region via Iceland but in that year it was discovered that a large number of birds return via Norway. It is still not certain where the knots wintering in West Africa go to breed. This knowledge is of more than purely scientific interest. Information on migratory routes is essential for conservation of a species that is known to be declining.

PARTIAL MIGRATION

The picture of migration is complicated by the movements of birds which give the impression of being resident in one place all year. It takes close observation of flights of birds, the sudden influx of large numbers, or the recovery of ringed individuals, to show that the population is shifting, with some local birds flying away for the winter or foreigners arriving to temporarily swell the population. Sometimes both processes may happen at once, the ebb of birds southward in autumn being replaced immediately by a flow from the north, with only a few local birds staying put.

Sometimes there is a distinguishing character that allows bird-watchers to identify aliens, such as the different subspecies of fox sparrow in various parts of North America or the northern subspecies of the long-tailed tit (which has an all-white head and no black eyestripe) that occasionally turns up in Britain. There may be differences in behaviour, too, as in the chaffinch. Bigger and paler continental birds visiting the British Isles in winter feed and roost in large flocks, mainly in open fields, while the native British chaffinches feed in small flocks, mostly near woods and hedges, and often roost singly in their breeding territories.

Partial migrations, in which only some individuals migrate, represent a shifting in the 'centre of gravity' of the species' range, and investigation of the extent of the movement throws light on the merits and disadvantages of the migratory habit. For instance, in some species there is a variation in the tendency to migrate, ranging from northern populations in which all individuals migrate to southern populations which are wholly resident. There are also differences in one population from year to year, in which movements away from the summer home are more complete in harder winters. In such species as starlings in Europe and northern mockingbirds in America, individual birds may migrate in one year and stay put the following winter.

BELOW *A flock of chaffinches feeds on spilt grain. The proportion of brighter males to drabber females tends to reflect the origin of the birds. In northern countries, more females migrate than males.*

The advantage of remaining in the north is that it cuts out long journeys in two directions and, because the days lengthen sooner in spring in northern latitudes, birds will come into breeding condition earlier. These facts reinforce the point that although migration enables the birds to escape the difficulties imposed by cold weather, there is a balance between the advantages of staying put and flying south. Dark-eyed juncos are popularly known as snowbirds because they appear in winter in the southern United States. Further north, in the New England states, the numbers that remain behind are inversely proportional to the severity of the weather. Cold fronts sweeping across the country in December and January drive the juncos southward and only a few remain behind to eke out a living around bird-feeders. Observations on captive

ABOVE The migratory habits of dark-eyed juncos depend on the weather. It is an advantage to remain near the breeding ground but bad weather and a shortage of food push many of the juncos south.

birds show that their urge to migrate (see page 60) is much greater when food is in short supply.

While partial migrants are influenced by weather, their behaviour is at least partly controlled genetically. From a French population of blackcaps in which three-quarters of the birds were migratory, researchers were able to selectively breed wholly migratory or wholly non-migratory groups within a few generations. In the wild, the balance is maintained because the migrants survive best in cold winters and the non-migrants in mild winters.

DIFFERENTIAL MIGRATION

The eighteenth-century Swedish naturalist Carl von Linné, (usually known by his Latinized name Linnaeus), who invented the modern scientific system of classifying animals and plants, gave the name *Fringilla coelebs* – the 'bachelor finch' – to the chaffinch because winter flocks in his native Sweden consist only of males. The females migrate south and Linnaeus' contemporary, Gilbert White, noted winter flocks in England that consisted almost entirely of females. The same pattern is seen in other partial migrants, with juveniles also joining the exodus. In dark-eyed juncos, three-quarters of the population wintering in Texas are females.

One possible explanation for this interesting phenomenon is that males are the dominant sex and drive the females and young from the best feeding places, so that migration becomes the better option. Another is related to size. Males are usually the larger sex and while this may be linked to their dominant status, it is also true that larger birds survive cold weather better because they can store more fat to fend off starvation. In addition, a bulky body loses less heat than a lean one.

The latter explanation may be more important, at least in the house finch. In western North America this species is not migratory, but the eastern population, descended from captives released in the New York area in 1940, has evolved a partial migration. Unusually, female house finches are dominant, yet they are the migratory sex. It seems that the males' larger size, or some other physical or physiological attribute, increases their chances of survival in cold weather. In support of this theory, in almost all birds-of-prey and a few waders it is the females that are the larger sex and they winter further north than the males.

Whatever the explanation for the difference in migration behaviour between the sexes, it is an advantage for male birds to remain as residents through the winter because this gives them a head start over rivals in staking out the best territories for the next breeding season. In wholly migratory species the males rush back to the breeding grounds in spring ahead of the females for the same reason (see page 147). Any male that managed to remain on or near its territory would have an even greater advantage. The situation is the same with female phalaropes, which take the initiative in courtship and may mate with more than one male. Although they do not hold territories, early arrival on the breeding grounds gives them a better chance of securing extra mates.

BELOW *The females of the polyandrous red-necked phalarope are larger and winter further north than the males.*

EUROPEAN ROBINS

For the British, the robin is both the national bird and a traditional symbol of Christmas. Compared with continental robins, which are skulking forest dwellers, British robins are confiding garden birds that are regular visitors to bird-tables and close followers of the gardener's activities that often give extra pleasure when they become hand-tame. Not only are they present throughout the winter but the male, and often the female as well, defends a territory during this season. Both sexes sing in defence of territories and their liquid notes further endear them to their human neighbours, who like to speak of 'their robins' and regard them with as much affection as they would a pet.

This is a picture of a resident that spends its life in one place – but, in fact, the European robin is a good example of

Many robins in western Europe are resident throughout the year and their sedentary habits hide the migration of a large number of individuals, mainly females, to winter in southern Europe and even North Africa.

a partial migrant. In the British Isles almost all robins are residents but some, especially those from south-east England, move to France and the Iberian peninsula for the winter. This is despite the fact that southern England has relatively mild winters. On the continent of Europe there is much more migration. Nearly all the robins in Scandinavia and Finland emigrate and pass through central Europe, heading for the Mediterranean, where Aristotle once thought they were redstarts in disguise (see page 13)! Bird observatories report significant movements, appearing after a night flight, across seas and through mountain passes. Some of these northern robins cross the North Sea and the British population is temporarily swelled until they move on.

In Belgium about half the robin population is migratory and the residents are joined by about the same number of immigrants from the north. There are significant differences between the behaviour of the residents and that of the migrants which winter in southern Europe. The migrant robins prepare for the autumn journey by moulting their feathers about two weeks earlier than the resident robins, and they then put on a considerable amount of weight.

In periods with average winters, the resident robins have much higher chances of survival from one year to the next. Even in the occasional exceptionally cold winter the death rate of residents is little different from that of migrants. This suggests that, overall, migration is the more dangerous option. Residents also have better breeding success. Resident males have the pick of territories – in gardens, where there is more food and shelter, rather than in woodlands. This gives them a better chance of attracting a mate from the females which arrive back in spring and the high-quality territory helps them rear larger broods. They can also extend their breeding season into the period when the migrants are preoccupied with moulting and fattening.

The obvious question is why any male Belgian robin should migrate when staying at home is such an advantage. The probable answer is that it is an option that is retained for use by birds that fail to settle in the best habitat in autumn. Their chances of survival if they spent the winter in marginal environments may be so reduced that migration becomes preferable. This seems to be the case for females which, as is usual with partial migrants, are more likely to move out for the winter. Migrant British robins are most likely to be female and in Belgium very few females are residents. Females seem to be less able to compete with the males for the best territories and, as they do not need to claim a territory in spring but instead choose a male with a good one, the extra risks of the migratory journey are worth taking.

LOOP MIGRATION

Some of the most famous sites for watching bird migration are worth visiting in both spring and autumn. Falsterbo in southern Sweden, and Gibraltar and the Bosporus at opposite ends of the Mediterranean, see streams of birds going in either direction according to season. By contrast, if you visit the American sites of Hawk Mountain in spring or Delaware Bay in autumn there will be little to see. Migrant birds do not always shuttle to and fro along the same route. Sometimes they have good reasons for taking alternative itineraries.

The solar radiation that warms the Earth's surface sets up a global circulation of the atmosphere which is distorted by the Earth's daily rotation to create the world's main wind systems. In temperate regions, between latitudes 30 and 60 degrees in both hemispheres, there are belts of westerlies. (Winds are always named after the direction from which they come.) Between these belts and the equator are the two sets of trade winds – north-easterlies in the northern hemisphere and south-easterlies in the southern hemisphere. Superimposed on this broad scheme are local effects that may vary through the year. While subject to short-term changes (winds from any direction may be encountered at any time in the westerly belts), the basic pattern is sufficiently constant for birds to have evolved migration routes that take advantage of them.

This is seen at its greatest extent in the so called 'loop' migrations of oceanic seabirds that travel in huge arcs during the time that they are absent from their breeding grounds. The map on this page shows the migration of the short-tailed shearwater of the Pacific Ocean; Manx and great shearwaters perform rather similar loops in the Atlantic. The Arctic tern's

Outside the breeding season, the short-tailed shearwater moves around the Pacific Ocean in a huge loop, making use of prevailing winds to make travel economical.

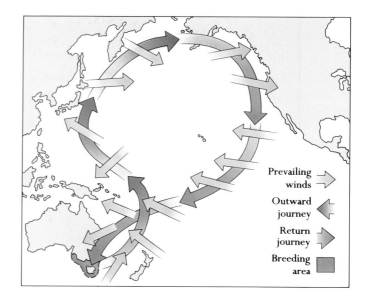

Prevailing winds

Outward journey

Return journey

Breeding area

record-breaking long-distance flight (see page 82–83) is also affected by prevailing winds.

European migrants are helped or hindered, according to season, by trade winds when crossing the Sahara (see page 80). They have little option in their route across the desert, although the red-backed shrike heads south through Sudan and Zaire in autumn and returns farther east through East Africa and the Arabian Peninsula in spring. In the New World there is a radical difference in spring and autumn migrations of many birds. The virtually extinct Eskimo curlew (see page 44) is one of several species that move down to the Atlantic seaboard in the autumn and launch themselves on a non-stop ocean crossing that takes them past the Bahamas and down to the coasts of South America. One advantage is that this route takes them first to Labrador, Nova Scotia and New England, where they feed on ripening berries to put on fat to provide fuel for the next stage of the journey.

Another advantage of this route is that it is one-third shorter than the alternatives: island-hopping through the Caribbean islands, crossing the Gulf of Mexico to the Yucatan Peninsula or travelling down the spine of Central America. However, it requires a very long non-stop flight and is used mainly by waders which are champion long-distance fliers. The Hudsonian godwit probably covers 4,500 kilometres (2,800 miles) in a three-day flight from Canada to South America. It is also likely that the diminutive blackpoll warbler follows this route to northern South America, although this has been disputed.

The third advantage of the ocean route is that it enables the migrants to take advantage of the prevailing north-westerly winds to carry them out to sea in the direction of Bermuda. They then keep going until they meet the north-east trade winds, which blow them towards the Lesser Antilles and South America. In spring, however, when the birds are heading northwards, these winds are a hindrance. Furthermore, the land is not ready to receive them, because spring comes late to the Atlantic seaboard as a result of the influence of the cold Labrador Current. Nevertheless, by flying northwards through the middle of the continent, the birds find land that is already frost-free and capable of supplying them with food at stopping points.

Climate is also the reason underlying the loop migration of the Russian population of the black-throated diver (known in North America as the Arctic loon). In autumn it flies south via the Black Sea to southern Europe, but its return journey in spring is north-west to the Baltic Sea and then east back to its breeding ground. The Baltic thaws much earlier, under the influence of the Gulf Stream, than do waters in central Eurasia, which makes the westward loop a more attractive option for the return trip.

The sooty shearwater breeds in two areas: New Zealand and adjacent parts of Australia, and southernmost South America and the Falkland Islands. Both populations migrate around the Pacific and Atlantic Oceans, following routes based on the prevailing winds.

LEAPFROG MIGRATION

In 1920 the American ornithologist H. S. Swarth published a classic study of the fox sparrow, a bird which exists as several distinctively coloured subspecies living across North America. The most widespread reddish (fox-coloured) subspecies nests in northern Canada and Alaska and migrates to New England and the southern states. On the western coast there is a complex of differently coloured subspecies which exhibit what has become known as 'leapfrog' migration. These different forms maintain separate nesting and wintering grounds and the northernmost subspecies, nesting in Alaska, migrate furthest, leapfrogging over more southerly subspecies which do not move so far.

Leapfrog migration occurs in a number of birds, such as kestrels, Canada geese and waders. In western Europe, the

Migrations of different populations of the fox sparrow on the west coast of North America are complex. It is not known why northern populations migrate further south than their southern neighbours.

ringed plover provides a good example. The British population is resident; Danish birds migrate to south-west Europe; those nesting in southern Sweden and Norway fly to West Africa; and the high-Arctic population of northern Scandinavia and Russia flies the furthest, reaching southern Africa.

Leapfrog migration has defied explanation, although several theories for it have been put forward. It is an obvious advantage for British ringed plovers to remain as near as possible to their breeding grounds. They avoid the cost and danger of migration and they can start breeding as early as possible in the spring. Northern birds have to escape a harsher winter but it is not clear why they need to travel so far. One suggestion is that they need to avoid competition with other populations, but Alaskan fox sparrows settle among resident Californian relatives. It will probably require considerable research into the details of the costs and benefits of migratory and residential lifestyles to finally solve the problem of leapfrog migration.

IRRUPTIONS

In the course of this year, about the fruit season, there appeared, . . . some remarkable birds which had never before been seen in England, . . . which ate the kernel of the fruit and nothing else, whereby the trees were fruitless . . . The beaks of these birds were crossed, so that by this means they opened fruit as if with pincers or a knife.

MATTHEW PARIS, 1251

This observation by an English Benedictine monk and chronicler is one of the first records of an irruption of crossbills. An irruption is an erratic form of migration in which birds move out of their normal range in such numbers that they often excite the attention of the general public as well as bird-watchers.

Crossbills live in the coniferous forests of both the Old and New Worlds, where they use their unique crossed bills for extracting seeds from cones. Good crops of cones are inter-mittent, and when they fail crossbills are forced to leave their northern forest homes and wander in search of alternative sources of seeds. Places well south of their normal range are invaded by large numbers of crossbills.

Species of birds that 'irrupt' in this manner rely on a limited diet of one or two foods that vary considerably from year to year. They include finches, nutcrackers and, sometimes, tits and jays, all of which rely on tree seeds, and owls and hawks that feed mainly on small mammals. *Irruptions** of snowy owls and rough-legged buzzards (known as rough-legged hawks in North America) coincide with the 4-year cycles of population growth and decline of lemmings and other rodents, while great horned owls and goshawks irrupt less frequently in line with the 10-year cycles of grouse and hares. If they cannot find their normal food, these wanderers will turn to novel food, such as the apple pips eaten by crossbills.

* Strictly speaking, an *irruption* is an invasion of birds and their departure from their homeland is an *eruption*. The whole movement is, however, usually described as an irruption.

ABOVE *Pine siskins are familiar North American garden birds.*
BELOW *Project FeederWatch, involving 8,000 volunteers, showed how pine siskins irrupt in winters when seed crops fail in the north.*

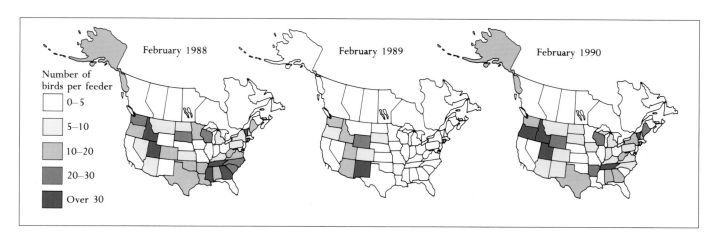

Irruptions lack the direction seen in regular migrations and it is not known whether the birds attempt to navigate in a particular direction like regular migrants. The birds head south on a wide band of bearings and individuals may be found far apart in successive winters. A siskin ringed in Germany one winter was found 2,200 kilometres (1,370 miles) to the east in Russia the next, and a cedar waxwing ringed in California was recovered later in Alabama, 3,000 kilometres (1,860 miles)

ABOVE *Fluctuating crops of berries force waxwings to lead a nomadic life in winter. The sudden arrival of a flock of these colourful birds transforms the dead branches of trees and hedges.*

away. Of 17,000 evening grosbeaks ringed at a site in Pennsylvania, only 48 were recovered there in subsequent winters, while 451 turned up over an area covering 17 American states and four Canadian provinces.

MOVEMENTS CAUSED BY WEATHER

Irruptions are irregular movements of birds caused by fluctuating food supplies. Their scale may be increased by the weather but meteorological conditions alone are enough to force birds to move in winter. Such traffic falls outside the definition of migration as a regular movement but overlays the migratory passage, intensifying and extending the flow of birds to a considerable extent.

Birds that normally survive the winter without long-distance migrations may be forced to move when cold weather locks up their food supplies. Waterbirds have to move in search of open water in which to dabble or dive, waders and others that probe soil and mud cannot feed in frozen ground, and seedeaters face starvation when snow blankets the ground. If the birds are lucky, they may beat storms and contrary winds and find a safe haven, but there can be heavy mortality if the wanderers are unable to find food.

Such movements are generally towards southern latitudes but the British Isles offer a haven to birds from continental Europe. When Europe is locked in the grip of frost, the Gulf Stream moderates the effects of the bitter weather on the British Isles and many birds trek westwards. If the cold continues and intensifies, the birds push across Britain and into Ireland, some even venturing into the Atlantic where a few are occasionally picked up by the wind and carried right across the ocean to North America.

Flocks of fieldfares appear and disappear with changes in the weather. Severe weather forces them to search for milder conditions.

CHAPTER 2

PREPARATIONS FOR THE JOURNEY

A bird must start to prepare for its migration long before setting off. Correct timing is essential. It must put on weight and perhaps moult into new plumage, so that it will be in peak condition for the journey. Finally, it must choose the right time and weather for its departure.

A small flock of dunlins roosting in winter quarters; their departure northward will depend on a number of factors.

GETTING READY

In 1702, Baron von Pernau, a pioneer student of bird behaviour, remarked: 'It is a very strange opinion if some believe that birds would emigrate, driven by hunger only. Instead they are usually fat when about to leave us'. Indeed, it was known to every man who netted or snared small birds for the market that the plumpest were those caught at the start of their migratory journeys.

Preparations for migration to winter quarters have to start many weeks previously. Throughout the summer months the bird is busy raising its offspring. This is a time of great activity and strain as the pair court, lay eggs and incubate them, and then, in many species, spend every day flying to and fro, collecting and delivering food to the growing family. By the time the adult bird is free of this responsibility it has lost weight and its plumage is frayed. It needs time for recuperation and renewal before its migration.

The period immediately after nesting is quiet. There is a noticeable absence of bird-song in the woods and along the hedgerows. The birds themselves are retiring. They are either feeding or resting. (The energy demands of moult mean resting is a good idea, and the birds are also hiding from predators which would be able to catch them more easily.) A few weeks later, there is a return to almost spring-like behaviour. Bird-song is heard again, although not at its full intensity. There are territorial disputes and even some courtship. Then the migrants disappear. Some, the failed breeders, for instance, will have slipped away a while ago: there are movements of waders as early as June. Many birds have gone before we realize they are no longer with us, but others are 'harbingers of winter' because we see the changes in behaviour which precede their departure.

The gatherings of swallows on telephone wires, for example, are a familiar sight in September and bird-watchers may note large numbers of swallows entering reedbeds to roost for the night, or warblers gathering to feed in the day (see page 44). Similarly, waders that have nested inland move to estuaries. There may be a slow drift of birds across country to these marshalling points before the final departure. Among these 'passage migrants' are warblers which sing for a few days in gardens where they were not heard in summer. W. H. Hudson described how shepherds on the South Downs, the range of rolling hills behind the south coast of England, used to trap large numbers of passage wheatears that gathered on the sheep-cropped turf from mid-July to the end of August. The birds were sold to be eaten by 'London stockbrokers, sporting men, and other gorgeous persons, accompanied by ladies with yellow hair', but the trade was eventually stopped by farmers objecting to the shepherds neglecting their proper duties.

The sight of flocks of swallows gathering on wires is a sign that they are getting ready for migration and will soon disappear.

JUVENILE EXPLORATIONS

When young birds leave the nest they usually spend some time with their parents and then disperse over the countryside to lead a nomadic life. Some of their wanderings are simply in search of food, or even to investigate possible breeding places where they will try to settle and raise a family when they are adult.

Ringing studies show that some young birds wander in any direction, gradually moving farther afield until their migratory journey starts. The extent of these wanderings was revealed by a study of the movements of British sand martins (known as bank swallows in North America) between the end of the breeding season and their migration to Africa. The Sand Martin Enquiry was set up after a team of bird-ringers trapped nearly 1,000 sand martins as they migrated along the river Lea near London and found that several had been ringed at distant colonies. As sand martins are social birds it was realized that there was the possibility of ringing and retrapping large numbers. From 1958 to 1968, 421,000 sand martins were ringed and 10,400 were recovered to provide evidence for wandering on a large scale.

After the nesting colonies have been deserted, sand martins roost at night in reedbeds. One roost studied was found to contain two million birds. Adults tend to fly south-eastwards through England, moving fairly quickly through the country as they prepare for their long flight to Africa, but they may stay at one particular roost for a week or more. Juveniles are more nomadic and wander between different roosts. Some sand martins hatched in southern England fly as far as Ireland but most movements are of a few kilometres only. Learning landmarks could be an important aim of these wanderings. Bird navigation is very accurate, as will be shown in Chapter 3, but

When this young lesser black-backed gull fledges, it will fly south, while its parents remain behind near the breeding colony.

landmarks are used for final recognition of the destination. In the following spring the returning birds need to find good feeding places, as well as nesting sites, as quickly as possible after their arrival. Prior knowledge would obviously be an immense advantage.

Juvenile sand martins ringed at a roost near Chichester were subsequently recovered in many parts of the British Isles before they migrated south to Africa. The size of the circles relates to the number of birds recovered at each site.

Chichester

FUEL FOR THE JOURNEY

Today, the Eskimo curlew is one of the world's rarest birds. This is in dramatic contrast to the situation a century ago, when huge flocks of these waders migrated down the north-eastern seaboard of North America. Waiting hunters shot them by the waggonload and in a few years the species had been all but wiped out. Eskimo curlews earned the name of 'doughbirds' because when they fell to the ground, their bodies split open to reveal a thick, dough-like layer of fat.

To acquire a store of fat, migrants must have an abundant supply of nutritious food so they can fatten up quickly and efficiently before it is time to leave. To find this food some birds change their diets or feeding habits, or gather at special feeding grounds. The flocks of Eskimo curlews used to fly across Canada and gather in Labrador, where they gorged on crops of ripening crowberries and bilberries.

The change of diet can be quite surprising. It is hard to imagine swallows feeding on anything but flying insects seized as they swoop through the air, yet the barn swallow of North America (the same species as the European swallow) turns to eating berries before the autumn migration.

A berry is a large, nutritious package. It is the plant's method of getting its seeds distributed because, with its juicy flesh and brightly coloured skin, it is, in effect, asking to be eaten. Animal food often comes in tiny, active packets, such as insects, which do their best not to be eaten by hiding or fleeing. When a bird is able to gorge on animal food, however, it can fatten quickly. This has been demonstrated in a comparison of the migration strategies of two similar birds – the reed warbler and the sedge warbler.

Both warblers migrate from Britain to tropical Africa but the sedge warbler puts on weight in southern England and flies to Africa probably in one hop, while the reed warbler stops to feed in Portugal. The reason for the difference appears, ultimately, to depend on the species' feeding behaviour. Both species feed by picking tiny aphids from vegetation, or by flying out to snap up hoverflies and bluebottles, but the sedge warbler is more adept at collecting small, slow-moving insects.

In late summer, British sedge warblers gather in reedbeds along the southern English coast, or in northern France, to feed on a short-lived glut of aphids swarming on the reeds. At best, a sedge warbler can double its weight in under three weeks. By the time they are ready to leave in August, the stocks of reed aphids are dwindling through south-western Europe, and the sedge warblers pass through without stopping. Reed warblers cannot feed on aphids so efficiently and opt to migrate in stages to Portugal, where they hunt for still abundant flying insects and put on weight for the shorter hop into Africa.

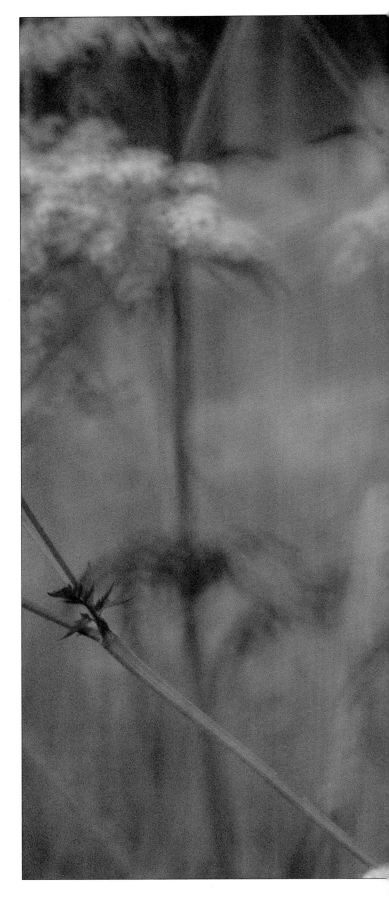

The reed warbler migrates to Africa in stages, so it does not need to lay down large amounts of fat as fuel.

THE MOULT

A bird's plumage gradually wears out. Apart from rubbing against foliage or the nest and the friction between neighbouring feathers, the plumage suffers from the forces imposed on the feathers during flight. If you look at a moulted feather you will see that its colours are faded and the edges are frayed. Sometimes the fabric of the vane is so threadbare that it is almost translucent. If it were not shed it would start to fall apart. The bird itself also has a shabby, worn appearance. This is a serious matter because the plumage is beginning to lose its insulation and flying becomes more strenuous. So the frayed and faded suit of feathers is replaced at least once a year.

However, the situation has to get worse before it gets better. Shedding the feathers leaves the bird in a temporarily poor state. When a moulting bird the size of a crow or a hawk flies over, the gaps in its wings can be clearly seen. Flight is laboured and less efficient, yet the bird must find the protein for manufacturing the feathers. It also requires energy for the process. There is very little information about whether birds use energy-rich food stored in their bodies before the moult, or whether they collect it during the moult. If the latter is true, they are further handicapped because their inefficient flight hampers foraging.

The effect of moulting on the energetics of flight has not been studied in detail, but there is plenty of evidence to show that birds follow the general rule of avoiding a moult when they have heavy demands on their resources. The demands are greatest when they are breeding and migrating. The corollary is that the moult is also often timed to eventually put the bird in optimum condition for a long flight.

Ornithologists study the progress of moult by examining the primary flight feathers – those at the tip of the wing. They are easy to scrutinize when a bird is being ringed (see pages 14–16) and their condition is critical for flight and hence migration. It takes a small songbird about 10 weeks to replace

its primary feathers and a wader about 15 weeks, but the time is very variable both between and within species.

A number of migratory birds compress the moult into a shorter period of time. For instance, among British finches, migrant species complete their moult in 8 to 11 weeks, compared with 12 weeks for non-migrant species. The moult may be so rapid that flight becomes difficult or even impossible. Snow buntings nesting in the Arctic have little time between nesting and the onset of winter. They moult so many flight feathers at once that they fly with reluctance and may become completely flightless for a few days – and they lose the tail altogether.

Complete flightlessness is a feature of the moult of almost all the wildfowl – the ducks, geese and swans – as well as some auks, such as guillemots (called murres in North America), and the grebes, cranes and rails. These birds must take refuge from predators on islands or on open water, or skulk in dense vegetation. Barnacle geese have a very rapid moult lasting five to six weeks and they are flightless for about 25 days.

Another solution to the problem is to start moulting while still breeding, as snow buntings do, but more commonly migrants delay the moult until they have arrived in winter quarters, or they have a suspended moult, which starts before departure, stops during the journey and is resumed after arrival. Delayed and suspended moults are quite common in waders nesting in the Arctic, where the summer is too short to complete the moult after breeding. As an example of both, over half the grey plovers (called black-bellied plovers in North America), wintering in Britain arrive before starting to moult, while the rest have only one to three new primary wing feathers. The latter may be failed breeders that had time to start moulting before departure from their Siberian breeding grounds. Once they have arrived, moult continues apace, except in a few of the birds that do not moult until spring, when they are preparing to fly back to the Arctic.

The disadvantage of delaying the moult is that part or all of the migratory flight is undertaken with substandard plumage, but the benefit is that the birds have plenty of time to moult when they reach their destinations. Dunlins moulting on Alaskan breeding grounds must hurry to complete their wing moult in 60 days, while those moulting in winter quarters to the south are able to take over twice as long. However, it is worth trying to fit in the moult between breeding and migrating if there is plenty of food. The lesser whitethroat, a European warbler, hunts insects but turns to fruits, such as blackberries, in the late summer. By gorging on this rich diet it can moult and fatten up for migration very quickly, and thus head for Africa in perfect condition.

The wing of this moulting starling has a mixture of new and old feathers. Most birds avoid migration while they undergo their moult.

RIGHT *The turnstone, like many other birds, has a distinctive breeding plumage (*ABOVE*). This is lost through a moult after nesting and a turnstone in winter (*BELOW*) is less boldly marked.*

MOULT MIGRATION

In the second half of summer huge flocks of shelducks gather around Heligoland in the Wadden Sea, the shallow waters in the angle formed by the Netherlands, Germany and Denmark. Numbering some 200,000 birds in total, they come from the British Isles, the Low Countries, Scandinavia and the Baltic. Their purpose is to moult, which leaves them temporarily flightless. After moulting, the flocks disperse and British shelducks return to the breeding grounds.

The strange fact about this 'moult migration' is that most of these shelducks are breeding birds and they have left their ducklings behind. The ducklings have banded into crèches of 100 or so, which are guarded by a few adults who remain behind and moult while on duty.

Similar moult migrations are undertaken by several species of geese and ducks, as well as members of other groups such as auks, divers (known as loons in North America) and flamingos. Their function seems to be to take the birds to a refuge, such as a lake or sheltered bay, where they are relatively safe from predators while flightless. Alternatively, or perhaps additionally, the moult migration takes them to productive feeding sites. This may be the explanation for the moult migration of pink-footed geese from Iceland to north-east Greenland. By flying north they find fresh, lush vegetation on the tundra, where the growing season starts later. In some species, in which non-breeders are involved in moult migration, a function may be to remove them from competition over food with the families.

Shelducks (and a single oystercatcher) roost on a mudbank. These boldly patterned ducks have the strange habit of leaving their young and flying long distances to special moulting places.

MOTIVE TO MOVE

'Yea, the stork in the heaven knoweth her appointed times.' So wrote the prophet Jeremiah. The timing of migration is crucial. A bird must be able to time the exact moment of departure to coincide with 'fair winds' on the way, and it must arrive at its destination at the appropriate time (not, for instance, when the ground is still snow-covered and food cannot be found).

Two sets of factors are at work to bring the bird into peak condition at the right time. There are long-term, ultimate factors programmed into the birds' behaviour that determine when they should arrive at their breeding grounds in spring or reach vital feeding grounds in autumn. These are affected by the annual changes in seasonal food supply. Short-term, proximate factors provide 'triggers' that initiate both the preparations for the journey and the actual departure. The triggers also come in two forms: an internal 'clock', and changes in the external environment. They can be compared to a domestic central heating system that has a clock to activate the system at a preset time, and a thermostat that brings it into operation only when the house becomes too cool for comfort.

The evidence that birds, and many other animals, have an internal clock for regulating their activity comes from experiments with captives. Birds that stop feeding and go to roost at the same time each day could be responding merely to the external stimulus of sunset, but experiments have demonstrated that this is not so. Birds kept in rooms with 24-hour, continuous light show the same cycle of activity and rest and go to roost when dusk would be falling outside. So roosting must be under the control of an internal clock.

However, the clock runs on a cycle of slightly less than 24 hours, so captive birds living in continuous light go to roost a little earlier each day. When the normal succession of night and day is resumed, the birds revert to a 24-hour cycle of activity because their internal clocks are continually being reset by external cues. In nature, the birds will always have sunrise and sunset to reset their clock and maintain a steady 24-hour rhythm of behaviour.

The internal clock is the ultimate control for all migratory behaviour, setting in train all the events of preparation and the journey itself, but it is modified by the environment – the weather, the bird's state of nutrition and social interactions with other birds. Which mechanism gets the upper hand and triggers the appropriate behaviour depends on circumstances. For instance, early in the departure period, environment wins and birds typically wait for fine weather and tail winds before departure (see page 53–54) Later, when there is a danger of leaving too late, internal control takes over and forces the birds to leave in bad weather.

The migration of the nutcracker is triggered by a failure of the crops of nuts and seeds that form its staple winter diet.

TIME TO GO

To see migrants set off, you have to be in the right place at the right time. Lucky birdwatchers may see flocks of waders taking off from estuaries and climbing steadily as they fly out to sea. Once the birds are on the move, however, the ideal place to be is in front of a radar set, where flocks of migrants show up as points of light that coalesce and track across the screen. Sometimes it looks as if a whole section of coastline is moving, as tens of thousands of birds take to the air simultaneously and head out to sea in the same direction, travelling on a broad front.

Without the benefit of radar, birdwatchers can get a false impression of what the birds are doing. A comparison of visual and radar observations of finches leaving Minsmere, on the south-east coast of England, showed how observers with binoculars can be completely misled. They noted most finches migrating southwards in autumn when the wind was from the south. The few that flew over when the wind was from the north, and apparently from a favourable quarter for carrying

migrants south, were heading into it and back towards their nesting grounds in an apparent 'reverse migration'. Radar showed the answer to this conundrum.

The true migration was taking place with tail winds and the birds moved on a broad front, flying in a southerly direction out to sea, but at an altitude too high to be seen through binoculars. When there was a head wind from the south, the birds avoided setting out to sea and flew low along the coast, so becoming visible to the watchers. The strange northward movement in the wrong direction took place at low altitude at the same time as the main flight was passing high overhead and out of sight. It is a phenomenon that has often been recorded, and probably consists of birds that have reached the coast with insufficient reserves for a sea crossing and are returning inland in search of a place to feed and roost until they are ready for their long migratory flight.

More difficult to explain is the British ornithologist David Lack's observation of a 'black snowstorm' of millions of swallows flying southwards up a valley in the Pyrenees in September. The wind was from the south but changed to blow from the north within the space of half an hour. Lack then saw some flocks turn round and fly back down the valley, heading away from their destination in Africa. Clouds settling over the mountains and filling the passes could have been a deterrent, because many observations have shown that birds avoid flying in clouds.

To study the departure of nocturnal migrants, the American ornithologist James Hebrard trained a searchlight over the tops of trees in a small wood on an island off the Louisiana coast. The birds had arrived across the Gulf of Mexico previously and were heading north to their nesting places. About an hour before sunset, the birds stopped feeding and became quiet. Between about 40 and 70 minutes after sunset, they took off singly and without calling. Most climbed steeply in a northerly direction, but if there was a northerly wind blowing, some took off into it and turned when they had gained height.

The departure of waders from a mudflat is easier to watch because it takes place before sunset. It also follows a different pattern. During the afternoon the birds start to call noisily and small flocks take off and circle, then tumble back to the ground. Sometimes, it appears that they have decided to leave and the bunched flock spaces out into a V or an echelon formation for a sustained flight, but the decision is rescinded and they come back. Eventually, through some joint agreement, they do set off, settle once more into formation and climb steadily until lost from sight. The communal activity and decision-making are needed if the flock is to fly together and gain the benefit of formation-flying (see page 95). Perhaps it also enables the birds to make joint decisions about navigation, which help ensure that they achieve a more accurate landfall.

Pink-footed geese prefer to set off when the wind is blowing from a favourable direction, but they may be forced to move in adverse weather if they are running out of food.

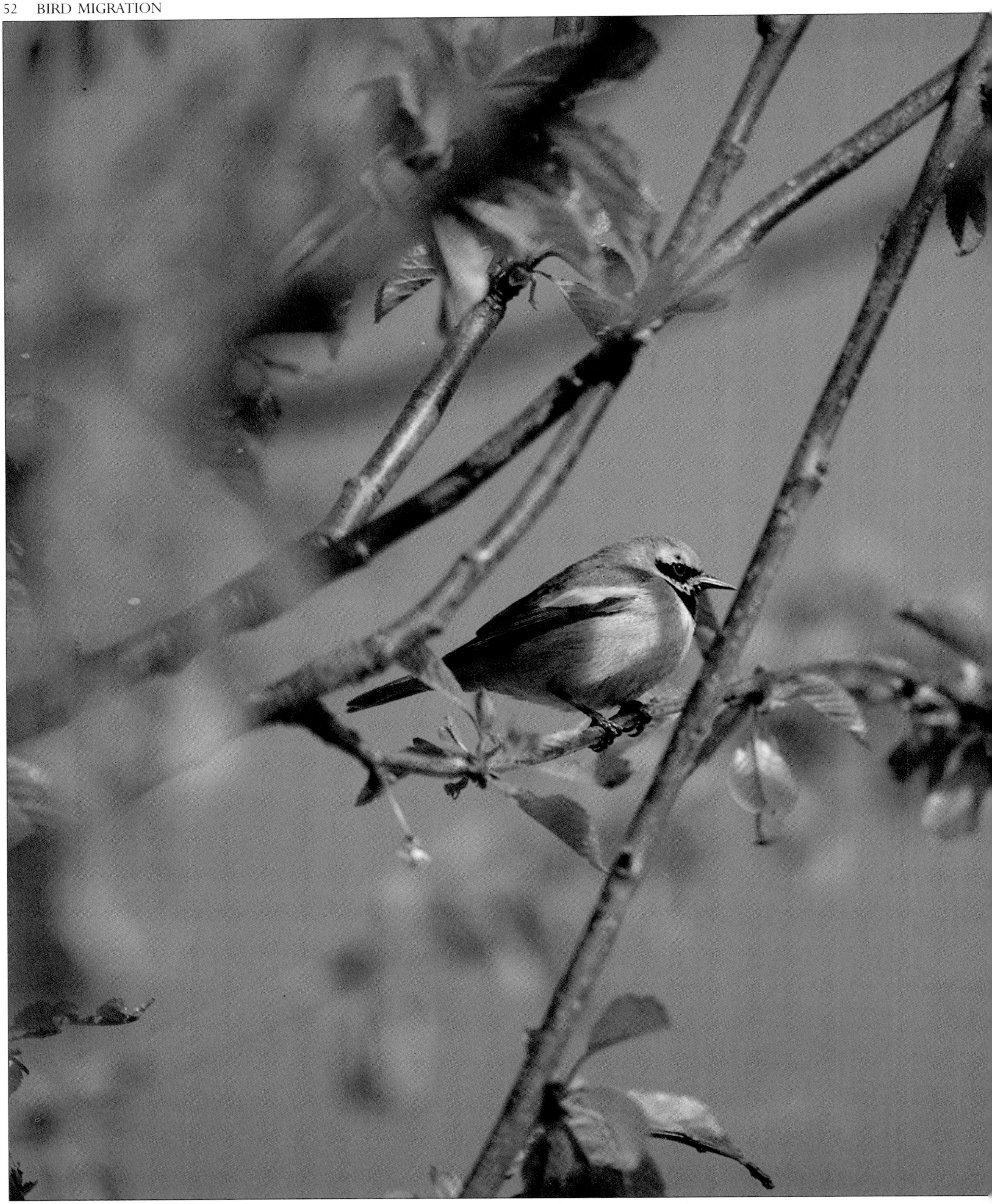

FAIR WINDS

William Bartram, an eighteenth-century American botanist, wrote: 'I have seen vast flights of the house swallow (*hirundo pelasgia*) and bank martin (*hirundo riparia*) passing onward north toward Pennsylvania, where they breed in the spring . . . and like wise in the autumn in September or October, large flights on their return southward. And it is observable that they always avail themselves of the advantage of high and favorable winds . . . In the spring of the year the small birds of passage appear very suddenly in Pennsylvania, which is not a little surprising, and no less pleasing: at once the woods, the groves, and meads, are filled with their melody, as if they dropped down from the skies. The reason or probable cause is their setting off with high and fair winds from the southward; for a strong south and southwest wind about the beginning of April never fails bringing millions of these welcome visitors.'

It is a common observation, also, that departures are delayed by bad weather. Rain and fog keep the birds grounded, perhaps because they cannot use their 'sun and star compasses' (see pages 63–65) for steering, or because flying is difficult with wet feathers. Head winds hold up migration for the obvious reason that they can slow the rate of passage to the dangerous extent of causing the birds to run out of fuel. In adverse weather swarms of birds may be held up. If they have been on passage, there are spectacular 'falls' of birds (see pages 108–109) which are released suddenly back into the air when conditions improve.

During the latter half of April, the population of pink-footed and greylag geese that have overwintered in Scotland and England start their return flight to breeding grounds in Iceland and north-east Greenland. They make their way northwards to the northern coasts of Scotland. I have seen flocks of one or other of these geese gathering at Cape Wrath, at the far north-west tip of Britain. (From a distance and in bad weather, the two species cannot be distinguished.) They streamed overhead, swept by a southerly gale over the open sea. Then the flocks checked and began to break up. Small parties detached themselves, lost altitude, turned and came back over the waves, beating their way into the teeth of the gale. The wind had seemed set to sweep them to their goal across the North Atlantic but something was putting the birds off, and the fields behind the cliffs became a transit camp for gaggles of geese.

Despite the apparent assistance of the southerly gale, the weather was probably too risky for the geese to make a long sea crossing. They prefer a clear atmosphere so they can see where they are going, and settled weather which will not suddenly subject them to contrary winds. Although they like

Countless migrant birds, such as this golden-winged warbler, filter south through North America in autumn. They may pause in one place until a weather system carries them onwards.

Above *A perfect day for flying. Barnacle geese prefer to migrate in clear weather and with fair winds.*

tail winds to blow them along, geese try to avoid very strong wind from behind, perhaps because they may get blown off course unless it is directly behind them.

The best weather to start migration is a combination of clear skies, to aid the birds' navigation and keep them dry, and tail winds, to blow them along. These conditions form in different meteorological situations. In anticyclonic, high-pressure weather, the atmosphere is clear and winds are light, so birds will not drift far even when flying into head winds. For the geese travelling from Scotland to Iceland, the best conditions occur when a ridge of high pressure reaches down from Greenland and a clockwise circulation carries them north-west towards Iceland.

The most impressive link between weather and the start of migration occurs off the coast of eastern North America, between Nova Scotia and Virginia, where millions of waders and songbirds collect as the weather deteriorates after the northern summer. They are heading down the coast into South America and the shortest route is over the ocean. When a cold front passes over, the coastal area empties of hordes of birds which disappear out to sea. They are taking advantage of the north-westerly winds that follow the front and they are swept along until the front slows down and weakens. The winds and rain slacken to allow the birds to fly through the front until they are picked up by the north-east trade winds and carried to Bermuda and beyond, into South America. They are not always successful. Sometimes the front does not weaken and many birds are drowned or struggle back to land, while a few are swept across the Atlantic to make landfall in Europe (see pages 108–109).

Birds cannot always choose when to set off and, as the migration season progresses, their options are reduced because they are running out of time, and they may eventually be forced to leave in inauspicious circumstances. Geese and cranes, for instance, which migrate in a series of relatively brief journeys with frequent stops, often do not leave a staging point until food runs short, and they may be caught out by bad weather. There is a well-documented example of this involving a group of pink-footed geese, which had to leave Iceland when heavy snow smothered their feeding grounds. They were forced to fly in strong westerlies, which drove them off their course to Scotland. Some turned and beat their way upwind to the Scottish coast, but others landed in Norway and Denmark and had to fly back across the North Sea when the wind changed.

WHEN TO FLY: NIGHT OR DAY?

A bird travelling in short stages has the choice of flying by night or by day. Only those making long-haul flights face flying through both day and night, although they can still choose when to take off. The majority of birds can be classed either as 'diurnal' (daytime) or 'nocturnal' (night-time) migrants (although probably no species is exclusively one or the other), and their choice of flight time is an important, although poorly understood, part of their migration strategy.

Many small birds, including warblers, thrushes, American sparrows, cuckoos, wrens and flycatchers, are nocturnal migrants. They typically go to roost around dusk as usual but then re-emerge and take off within about half-an-hour after

The white stork migrates by day because it relies on thermals for economical flight. These rising columns of warm air form only when the sun is shining and the ground has warmed up sufficiently.

sunset, when twilight is giving way to darkness and the stars are coming out. Although the migrants do not travel in dense flocks, the movement is well synchronized and masses of birds depart in the space of a few minutes. The birds keep going through the hours of darkness and, if they are not over water, land at dawn.

Diurnal migrants set off before sunrise, sometimes too early to be seen by human observers but light enough for the birds to avoid obstacles until they have gained height. They fly for usually no more than a few hours before they stop to feed. These birds are mostly seed-eaters and fruit-eaters such as finches, buntings, pipits, pigeons and starlings, but also include insect-eating swallows, martins and bee-eaters which can feed on the wing as they travel.

Perhaps the most interesting point about nocturnal migration is that it is the strategy of birds that are normally diurnal in their habits. There is no clear reason why so many birds should prefer to migrate by night. Using the setting sun to orientate and then maintain a course with reference to fixed star patterns may be easier than continually compensating for the movements of the sun. Avoiding predatory birds is important, and another advantage may be that the night air is more favourable for flight. It is cooler and more humid, which reduces the stress of vigorous exercise; it is denser, which makes flying easier; and it is less turbulent. One further advantage is that night-fliers have the entire day for feeding. It is therefore strange that other birds should fly by day, especially if it means that they fly for only a few hours before stopping to feed.

There is no doubt about the reason why one group of birds migrates by day. These are the large birds – birds of prey, herons, pelicans, cranes and storks – that make use of thermals (rising currents of warm air) to help them make their journeys. As will be described on pages 93–95, soaring in thermals is very economical of energy, but it is dependent on the birds finding suitable weather conditions. This is not merely a matter of the sun warming the ground. The best thermals form when there is cool air over a warm surface; this occurs after the passage of a cold front. Travel time is short because it is linked to the daily warming of the ground. The birds have to wait until mid-morning before they can set off, and must land again when the sun loses its strength in late afternoon. One exception to this rule is the observations of white storks using warm air rising above gas burn-offs at oil-drilling sites in the Sahara. Flocks have been seen circling above the flames on a dark night, long after natural thermals had disappeared, and continuing their journey northwards. With oil-wells spaced out across the desert, it must be possible for the storks, and perhaps other soaring birds as well, to speed parts of their crossing of this inhospitable region in this interesting way.

A flock of starlings settles into its roost. Starlings travel by day and rely on a sun compass for navigation. They take off in the early morning but must stop to feed during the day.

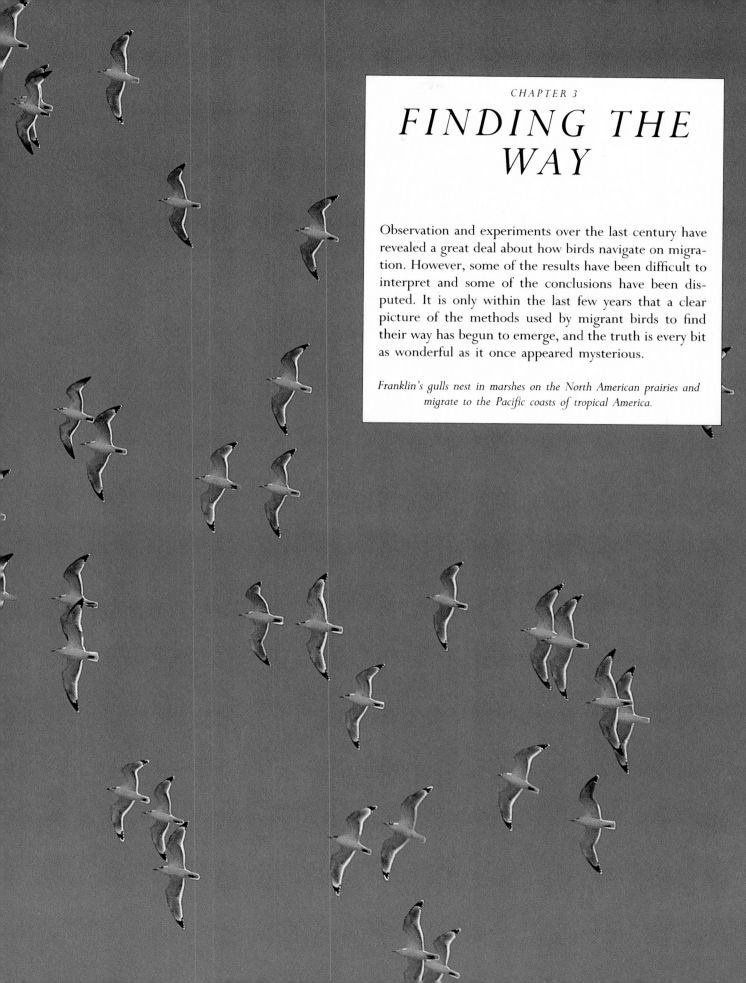

CHAPTER 3

FINDING THE WAY

Observation and experiments over the last century have revealed a great deal about how birds navigate on migration. However, some of the results have been difficult to interpret and some of the conclusions have been disputed. It is only within the last few years that a clear picture of the methods used by migrant birds to find their way has begun to emerge, and the truth is every bit as wonderful as it once appeared mysterious.

Franklin's gulls nest in marshes on the North American prairies and migrate to the Pacific coasts of tropical America.

STUDYING NAVIGATION

Imagine standing on the deck of a ship, several days out to sea, and watching a land bird flying past and disappearing towards the horizon on an undeviating course. How does it know that it is flying in the right direction? Or imagine finding a swallow at its nest under the eaves, bearing a numbered ring that shows that it was captured at the same spot last year. How has it found its way back with such precision? These questions have formed the traditional 'mystery of migration', the riddle being: how on earth do most birds find their way with such unerring accuracy?

An often quoted example of the mystery of bird navigation is the juvenile European cuckoo that is not only reared by foster parents of an alien species, but also embarks on its first long flight to southern Africa a month or more after its true parents have already departed. The young cuckoo must have acquired all the information it requires for navigation from Europe to Africa by inheritance through its parents' genes, and must therefore have an instinctive knowledge of its route and destination.

The complete story of how birds find their way has still to be worked out in all its details, but 30 years of research have provided some interesting clues that at least reveal the amazing sensory capabilities of birds, and the ways these can be used for route finding. In any discussion of navigation, as with many other biological subjects, it must be remembered that scientists have studied only a few bird species. These have been birds which are easy to keep under observation, such as homing pigeons, starlings and indigo buntings. Using different species does not always help interpret experiments because the birds may be behaving in different ways. Problems of

OPPOSITE *Adult European cuckoo; the ability of the young to migrate across the world without ever having had contact with their parents was early evidence that birds could navigate by instinct.*

BELOW *Observations on free-flying mallards have helped to demonstrate how they use the sun for orientation.*

interpretation may also result from the use of different types of apparatus and experimental techniques. However, it has to be assumed that navigation systems which have been discovered in the few species that have been studied operate generally in all migrants, although different species may use them in rather different ways.

We might expect that a young barnacle goose crossing the North Atlantic from Greenland to Scotland with its parents is not using its senses in the same way as a young cuckoo travelling alone to Africa, or that a willow warbler crossing the Sahara at night is not responding to the same sensory clues as a swallow covering the same route by day. The most important point is to realize that a bird has a range of navigational aids and strategies at its disposal and will use whichever are appropriate.

It is only possible to get hints of how wild birds find their way on migration in natural conditions and they usually come from witnessing birds making mistakes. For instance, G. V. T. Matthews, a British ornithologist, was studying mallards' ability to orientate by the sun by releasing captive ducks and recording the direction in which they flew. On one overcast evening the sun was setting out of sight in the south-west when a break in the clouds to the north-west caused a red flush low on the horizon. The mallards apparently mistook this for the setting sun and flew north-east instead of the expected north-west. In other words, the apparent position of the setting sun had shifted 90 degrees and the mallards had adjusted their heading accordingly.

Our knowledge of bird navigation systems comes mostly from experiments on homing pigeons or captive birds of several wild species. As domestic animals, homing pigeons are easy to control and manipulate in experiments with sophisticated equipment. Homing pigeons are descended from rock doves, which are not migrants, but they are capable of returning home when released from distances of 1,000 kilometres (620 miles) or more. It is assumed that they use the same senses to find their way back to the loft in the same way as wild birds navigating on migration.

Experiments on captive birds have involved songbirds such as indigo buntings in North America and garden warblers in Europe. The experiments make use of a behavioural trait peculiar to birds that migrate at night. At the time of year that they are preparing for migration, they become restless, fluttering against the bars of their cages and whirring their wings. This behaviour is usually known by the German term *Zugunruhe*. Normally placid, captive birds exhibit this migratory restlessness by fluttering in the direction that they would head if they were free to fly. Manipulating environmental cues, such as the direction of light from the sun or the orientation of the magnetic field, and observing the effect on the direction of the birds' restlessness, has been a valuable technique in studying bird migration.

PILOTING, COMPASS COURSES AND NAVIGATION

There are three ways of getting from A to B. The first is piloting, in which we steer with reference to a series of familiar landmarks. The second is to steer on a compass course, using some form of reference to make sure we keep heading in a straight line. The third is true navigation, in which we calculate our position on the Earth's surface, compare it with the position of our destination and then steer a course between the two.

Each system has its drawbacks. Piloting is no use if the landmarks are obscured. A compass course is unreliable if we have been displaced, because we will end up travelling on a parallel course and so miss our destination. True navigation also becomes unreliable if the reference points for calculating our current position are obscured.

— PILOTING —

We find our way around our neighbourhood by familiar landmarks. They become so well-known that we can get home from the shops or station without consciously thinking about it. When we travel on a new route, we start by studying the map and checking route numbers and signs. As we get to know the route, we learn landmarks which tell us when to change direction without having to check the signs. It seems very likely that birds do the same sort of thing. While they are moving around the country after leaving the nest (see page 43), they must be learning the lie of the land ready for the time that they return from their first migration. Moreover, they will be learning landmarks on their first migration flight, which will presumably help them on their return.

The view from a height of 1,000–3,000 metres (3,300–9,900 feet) is marvellous, with the landscape laid out below and all the details of topography presented in finer detail than on the best map. Even at night the coastline, rivers, islands and hills will be visible, and the evidence from radar observations is that migrating birds are making use of such natural features. There are other types of landmarks, however.

One possibility is that birds use sounds as landmarks, although this is very much in the realms of conjecture. It has been suggested that birds may even hear their own calls echoing back from water surfaces. This has the support of an experiment in which recorded bird calls were broadcast from a balloon and the echoes were clearly picked up. The croaking of frogs in their spawning ponds has been heard from an altitude of 1,000 metres (3,300 feet), so there is no shortage of sounds that are potential cues for piloting.

Another idea is that birds can detect infra-sounds. These are sound waves below our range of hearing, with frequencies less than 10 hertz, or cycles per second. Birds can hear sound

A 'bird's-eye view' gives a panoramic view of landmarks for accurate navigation. This flock of sandhill cranes at their staging post on the Platte River, Nebraska, acts as a target for other cranes.

of 0.5 hertz. Infra-sound is generated by waves breaking on the shore and by wind humming over desert sands or swirling up mountains, and it carries hundreds of kilometres. Calculations show that birds could detect the direction of infra-sound sources by means of the Doppler effect. (This is the effect of a sound changing its frequency as the source moves relative to the listener. The changing notes of train whistles or police-car sirens are familiar examples.) One problem with sound landmarks is that they may be masked by extraneous noise: the birds have to cope with the 'cocktail party effect' of interesting sounds needing to be picked out from a loud background noise of wind in the trees, waves in the sea, rumbling cities and even the beating of their own wings.

Stranger even than the use of low-frequency sounds is the idea that birds could be capable of using smell landmarks. It used to be thought that birds have virtually no sense of smell, except in a few special cases such as kiwis probing in the soil for worms. Research is now showing that some birds' sense of smell is not so poor and that they can make use of it in a variety of situations. Experiments conducted in the mid-1970s by an Italian research team, led by Professor Floriano Papi at the University of Pisa, showed that homing pigeons can be trained to respond to wind from a particular direction with a specific scent: olive oil or turpentine.

Two groups of homing pigeons were kept in aviaries isolated from breezes carrying natural odours. On top of each aviary there were two glass-walled corridors through which air could be blown to simulate a breeze. One group of pigeons received a flow of air from the south with the scent of olive oil added, and from the north with turpentine added. The second group received the treatment in reverse.

All the pigeons were taken to a place due west of the aviaries and released with a drop of olive oil or turpentine rubbed on the beak as a powerful olfactory stimulus. The pigeons assumed, from the strength of the odour, that they were nearer the 'source' of the smell than their aviary and flew in the appropriate direction to reach the aviary; in other words, they were trying to fly away from the 'source'. Birds with olive oil on their bills flew to the north if they had received an olive-oil-laden 'breeze' from the south and *vice versa*. The 'turpentine' group behaved in a similar fashion. It was also shown that pigeons deprived of their sense of smell were not so good at navigation.

Some other researchers have failed to confirm these findings; possibly, not all pigeons find smell landmarks useful. Perhaps only birds living near a powerful, reliably consistent smell would include it in their suite of navigational aids.

— SUN COMPASS —

We usually think of a compass as a magnetic needle pointing north towards the magnetic pole, but sun and star compasses are used in certain circumstances – for instance, in the polar

regions where the magnetic compass is unreliable. In each case the compass uses a point of reference so the user can maintain a fixed course.

In the late 1940s the German ornithologist Gustav Kramer demonstrated that starlings can orientate by means of the sun. They were trained to search for food in a particular corner of their cage. They were shown to be fixing the direction by using the angle of the sun because, when the apparent position of the sun was altered by mirrors, the starlings changed the direction in which they were searching by the same amount.

The sun's position in the sky continually changes but its movements are predictable: the sun is always due south at local noon. So a sun compass must be geared to a clock which tells the user the time of day so that he can compensate for the sun's movement. Kramer's starlings accepted a bright light as an artificial sun and, as it was kept stationary, their choice of direction backed anticlockwise at the same rate as the sun normally moves clockwise.

Birds, like many other animals, have an 'internal clock'. The clock is set by the cycle of night and day and can be reset by subjecting the animal to a changed light/dark sequence. The possession of a sun compass can be tested by this 'clock-shifting'. One group of birds is kept in normal daylight and a second is given a light/dark cycle six hours out of step with the sun's rising and setting. When tested, the second group

Adélie penguins set out from their rookery to feed in the ocean. They will use a sun compass to find their way back home after their time at sea.

orientate 90 degrees from the correct heading. By this means a number of species (mallard, starling, white-throated sparrow and barred warbler among others) have been shown to have sun compasses, and it is reasonable to suppose that they may exist in all birds. (Adélie penguins were tested in another way. They were taken from their colony and released on a feature-less snowscape. While the sun shone, they trudged unerringly along a compass course, but they wandered erratically when it clouded over.)

A sun compass will not work at night, nor on heavily overcast days, but birds may still be able to steer under clouds that are not too thick. Their eyes are sensitive to ultra-violet light which penetrates cloud better than the wavelengths visible to us. (We cannot see ultra-violet light because it is absorbed by yellow pigments in our eyes. Birds lack this pigment.)

Birds may also make use of small breaks in the cloud cover because they are sensitive to polarized light. The sun's rays are polarized when they are scattered by molecules of air to create patterns of light that are invisible to us. (Polarization can be visualized by imagining waves coming in towards the seashore.

The 'Kramer cage' was used to show that birds orientate by the position of the sun. A captive starling was taught to search for food in one part of the cage and its orientation in relation to the direction of the sun (indicated by the dots on these diagrams) was then determined in a series of experiments.

TOP First, the starling orientated north-west in natural sunlight.

MIDDLE The apparent direction of the sun was altered through 90 degrees by a mirror and the starling changed its orientation by the same amount.

BOTTOM The apparent direction of the sun was reversed by another mirror and the starling followed.

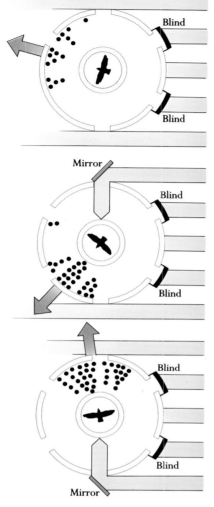

the axis does not move, the birds do not need an internal clock. They take their bearings from the fixed Pole Star and the pattern of constellations rotating closely around it.

This simple star compass becomes less effective for birds heading towards the equator, and beyond, because the Pole Star disappears below the horizon. Research shows that the indigo bunting, which does not migrate as far as the equator, uses the simple Pole Star compass but some European warblers, which cross the equator, may use a more complicated system, more like the sun compass, that involves an internal clock and moving stars visible in equatorial regions.

The moon is likely to be a less useful reference point than the sun because its course through the sky is complicated and there are nights when it is below the horizon. Nevertheless, some experiments with mallards show they can use a moon compass, in much the same way as the sun compass, and it is possible that this is true for many birds.

— MAGNETIC COMPASS —

The idea that birds have some form of magnetic compass was attractive to the pioneer investigators of migration. It would help explain how birds keep flying in a straight line in overcast or foggy weather. The problem was that early experiments failed to find any evidence for its existence. The first proof that birds are sensitive to the Earth's magnetic field eventually came from observations that robins in Frankfurt continued to show Zugunruhe (migratory restlessness) in a south-westerly direction – their normal migratory heading – even when confined to a darkened room with all cues from the sky removed. The researchers surrounded the cage with a large magnetic 'Helmholtz' coil which changed the field inside it. When magnetic north was shifted from its true position pointing to the magnetic pole, the robins changed the direction of their Zugunruhe fluttering accordingly.

This breakthrough was made when it was realized that birds detect magnetic fields only when they are moving through them. The original experiments had involved fastening Helmholtz coils or bar magnets to the heads of homing pigeons and were doomed to failure, in any case, because the sun compass is used in preference to a magnetic compass. Later experiments were performed on overcast days to demonstrate that pigeons use some sort of magnetic compass when they cannot see the sun. However, this does not register polarity – the direction of the magnetic north pole – like a compass needle swinging on its fulcrum. Instead, the birds are using a 'dip-compass'.

The Earth's magnetic field can be imagined as an immense bar magnet buried in the centre of the Earth. The lines of force, like those revealed by scattering iron filings around a magnet in the classic school science demonstration, radiate in huge arcs far into space and sweep back in at an angle, rather than running along the ground, like contours on a map. This angle is the angle of dip – the acute angle between the lines of force and the horizontal – and it varies over the Earth's surface, pointing downwards in the northern hemisphere and

The waves are polarized because they are travelling in one plane, along the surface of the sea. Unpolarized light would be represented by waves tilted at all angles between vertical and horizontal.) The angle of the plane of polarization depends on the angle at which light strikes air molecules, so the pattern of polarization in the sky depends on the position of the sun. A bird needs only a small patch of blue sky to determine the plane of polarization and work out the sun's position.

— STAR COMPASS —

Gustav Kramer was also the first to demonstrate that birds have a star compass. He showed that their Zugunruhe, or migratory restlessness (see page 60), is orientated under a starry sky but their movements become erratic when it clouds over. Later experiments in a planetarium showed how the birds were using the pattern of the stars.

If you keep watch on the night sky long enough you will see that the stars rotate slowly around an axis. (Strictly speaking, the stars are stationary and the Earth is spinning.) In the northern hemisphere a very useful cosmic coincidence has placed a single large star at this point: the Pole Star. Because

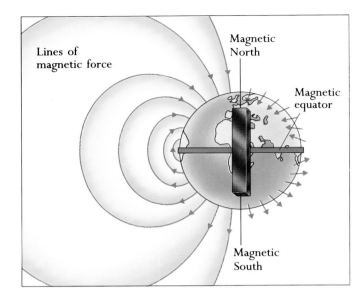

The Earth can be visualized as a huge bar magnet, with lines of magnetic force extending in great arcs far out into space. Birds are sensitive to the angle of dip, the angle at which the lines of force meet the Earth's surface, as indicated here by the arrows on the right of the Earth. The polarity of the magnetic field is shown by the direction of the arrows.

upwards in the southern hemisphere. At the magnetic poles, it is vertical and at the magnetic equator, horizontal. The magnetic compass of birds registers this angle of dip.

It is not yet known how the magnetic sense works. One theory is that the crystals of magnetite recently found in magnetic-sensitive bacteria and in the brains of some animals, including homing pigeons, make up the receptor. Magnetite is a form of iron oxide which, under the name of lodestone, was used for primitive compasses by mariners as early as the twelfth century. The suggestion is that there are bundles of magnetite crystals in the bird's head, whose response to the Earth's magnetic field when the head is moved triggers impulses in surrounding nerves. Another theory suggests that the magnetic sense is located in the back of the eye and that, because the bird's magnetic compass does not work well in pitch darkness, light must provide the necessary energy.

The bird's magnetic compass is an unreliable instrument and fails near the equator. Magnetic storms and anomalies in the Earth's field upset magnetic compasses and cause serious problems for human navigators. There is evidence that they also disrupt birds. Thomas Alerstam, the Swedish expert on bird migration, watched birds flying past Norberg in central Sweden, where huge deposits of magnetite and other iron ores create a magnetic field up to 60 per cent higher than normal. Some low-flying migrants landed and wandered around before taking off again; others suddenly dived and broke up their flock formations. The majority, however, were unaffected, perhaps because they were using another compass or were piloting by landmarks.

TRUE NAVIGATION

Successful migration requires that birds know how to adjust their heading to reach their destination if they have been blown off-course by winds too strong to maintain the original heading, or if they have become lost in bad weather. Compasses and landmarks are not enough because a displaced bird has to know in which direction its destination lies. True navigation is used by human travellers who mark their present position and destination on a map, draw a line between them and set off on the corresponding compass course. If a traveller is lost, he can take bearings to recalculate his position, draw a new line on the map and correct his course.

A pre-programmed compass and clock are sufficient for guiding a bird along its route, taking in changes of direction and perhaps following the continually changing heading of a Great Circle route (see pages 77–78), but the bird must know when it has arrived at its destination. A young bird returning to its nesting area will recognize landmarks that it learned during its first explorations (see page 70), and it will presumably do the same when it flies south for its second winter. It may be using landmarks that it has learned on its first journeys to check its position along most of the route.

The problem is more difficult for birds that have a choice of routes. Scandinavian chaffinches heading for the British Isles may take the direct route across the North Sea but some fly through Denmark and the Netherlands, heading first south, then south-south-west and then west-south-west. If the wind is behind them, they continue on the last course out to sea for England. Otherwise, they continue still further down the coast into France, and either turn west-north-west for a Calais–Dover crossing or turn south and west and make a northward crossing to southern England from Brittany. This route is three times the length of the direct crossing and involves changing course through 180 degrees. It would be interesting to know how the birds are programmed but, as diurnal migrants, chaffinches do not show *Zugunruhe*, or migratory restlessness, so cannot be tested like warblers (see page 60). It would be even more interesting to know how the direction and strength of the wind influences their choice of route.

For its first winter, a young migrant bird heads into the unknown aided only by its own internal guidance system, unless it is travelling with companions in a close-knit flock. Is it finding its way like the V-1, the original 'flying bomb' of the Second World War, that flew on a set course and cut its engine after the lapse of time which should have brought it over its target? Or is the naive bird also programmed to recognize the right habitat, not only for its final destination but for the intermediate staging sites?

Evidence that birds learn the position of their destination through experience and navigate towards it comes from a classic experiment performed by the Dutch ornithologist A.C. Perdeck. His subjects were starlings that migrate in autumn from the Baltic region on a south-west course through the Low Countries to northern France and the British Isles, as shown on the map on this page. Perdeck caught and

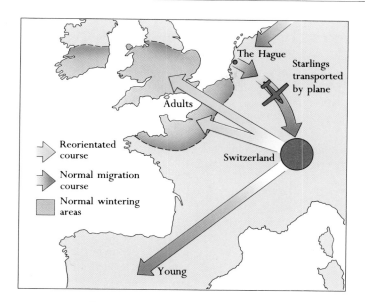

The Hague

Starlings
transported
by plane

Adults

Reorientated
course

Normal migration
course

Normal wintering
areas

Switzerland

Young

Taking migrating starlings from the Netherlands to Switzerland showed that only adults could compensate by altering course.

ringed many thousand starlings over several years as they passed through the Netherlands. Some were released immediately so that their recapture would show the natural wintering area but over 11,000 were put on board aeroplanes and released at airports in Switzerland, about 600 kilometres (370 miles) to the south-south-east.

There was a striking difference, after their release in Switzerland, between the behaviour of adult starlings, with at least one round-trip of migration behind them, and young birds on their first trip. No matter whether they were released singly or in flocks, the adults readjusted their course and headed north-west to be caught later on their normal wintering grounds. The inexperienced starlings continued on the same south-west course as they had been taking along the eastern coast of the North Sea. This led them into south-west France, northern Spain and occasionally as far as Portugal.

This is evidence that young birds merely fly on a fixed, presumably genetically coded, heading until some signal, either internal or from their environment, or a combination of the two, tells them that they have gone far enough. The adult birds evidently have a more sophisticated navigational system that informs them that they have been displaced and allows the mistake to be corrected. It must be based on their experience of their first winter because the displaced youngsters later returned to their new French and Spanish homes in their second winter. Most of them must have found, and learned, the position of a suitable place but a few returned to the traditional wintering area. Perhaps they had failed to find the required habitat in the first winter, and in default had returned to their original course in the hope of better luck in the second winter.

To find what happens on the return journey, Perdeck captured and ringed another 3,000 first-winter starlings on

their way north-east through the Netherlands in spring. They were also transported to Switzerland. The results suggest that they readjust and navigate back to their natal home around the Baltic Sea, although some settled in Switzerland.

The conclusions from Perdeck's experiment are what might be expected: inexperienced starlings head blindly on a compass course until they recognize, in some unknown way, that they have reached a suitable destination, but starlings that have already learned a destination can reorientate and navigate to find it after being taken off-course.

Reorientation is not practised by all birds, however. Some species are less attached to one area. In a similar experiment to Perdeck's, hooded crows were captured on their way north in spring in Lithuania and transported to Flensburg on the Danish border of Germany. Most adults and young birds maintained their original heading after release and settled in Denmark and southern Sweden; only a few adults reorientated and returned to their native home east of the Baltic Sea.

The behaviour of Perdeck's young starlings and the Baltic hooded crows is simple and easy to explain. If all goes well, the bird travels on a straightforward course between its summer and winter homes but it can be in serious trouble if it gets displaced, perhaps by a gale, or loses its orientation when its compasses are put out of action by thick cloud or magnetic anomalies. Many young birds are lost every year through navigational accidents, and they account for most of the bird-watcher's prize finds of rarities spotted far from their normal range (see pages 108–111).

Perdeck's experienced starlings show that birds know how to reorientate after they have been displaced and find their way back to their usual winter homes. This experiment is mirrored every year in real life when birds fly off-course. The way that they reorientate and find their way back to their destinations is the great unsolved mystery of migration.

Perdeck's starlings had two ways of getting back to their winter quarters. They could have a 'mental map', or grid, on which their destination is marked, and they would fly towards it, in much the same way that a human navigator is guided towards a map reference, by comparing his current position with the destination and steering to bring the two together. The navigator traditionally uses a sextant and chronometer to measure the position of the sun and calculate his position on the map. Despite a great deal of research there is no conclusive evidence that birds can achieve a similar feat. Although they can plot the sun's position with sufficient accuracy, they do not use their biological clock like a chronometer.

The second option is that displaced birds somehow record the direction in which they are being carried and work their way back to get on course. Despite some hints that pigeons transported in a distorted magnetic field were disorientated when released, experiments in which pigeons were drugged, deprived of natural light and odours, or given confusing magnetic information, have shown that the birds were still able to fly home. The conclusion has to be that, at the present state of our knowledge, this reorientation remains inexplicable.

PROGRAMMED ROUTES

Over 170 years ago, the German naturalist Johann Naumann realized that the time spent in *Zugunruhe*, or migratory restlessness (see page 60), is related to the distance a bird flies on migration. Later, it was suggested that the migratory urge as shown by *Zugunruhe* lasted for the same number of hours it would take an inexperienced bird on its first migration to reach its winter home. Even more precisely, it was found that the time that the captive bird whirrs its wings matches the calculated time it would spend in the air on the journey. This behaviour has enabled a research team at the Vogelwarte (Bird Observatory) Radolfzell in Möggingen, Germany, to gain some remarkable insights into the mechanism of migration.

German garden warblers migrate to Africa for the winter. Captive individuals change the mean direction of their orientation at the correct time, as shown by the concentration of dots in the circles (the latter represent the cages used in the experiments). Ringing recoveries show that free-living birds do indeed first fly south-west through Europe and then alter direction to migrate south through Africa (as shown by the arrows).

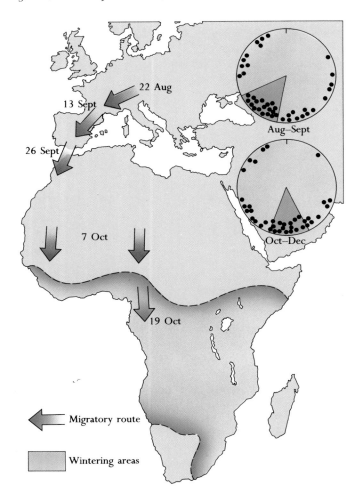

The willow warbler which leaves in August for tropical Africa shows *Zugunruhe* for at least 120 days, while the chiffchaff which sets out later and spends the winter around the Mediterranean and North Africa shows a peak of *Zugunruhe* two months later, and it lasts for only 80 days. This is support for the idea that young, inexperienced birds fly for a set time that would bring them to a pre-ordained destination, from which they wander in search of a favourable place to settle.

Another remarkable feature of these bouts of restlessness is that their orientation also mimics the changes in heading that free-living birds would be making on their journey. Captive garden warblers orientate their restlessness towards the south-west from late August to late September, the period when they would normally be migrating from southern Germany through France, Spain and into North Africa. Then, it swings to the south on the heading that the free-living birds will be taking across the Sahara and into the savannah regions on the far side. This finding agrees with the route worked out from recoveries of garden warblers ringed in Germany and is more evidence that the migratory journey of inexperienced birds is guided by pre-programmed, instinctive instructions.

Support for the inheritance theory is also given by cross-breeding experiments with captive blackcaps. German blackcaps migrate south-west and spend the winter in the western Mediterranean, while Austrian blackcaps head for eastern Africa. This involves a change of heading, similar to that of the garden warbler, in which the birds fly first south-east, then south. The researchers interbred German and Austrian blackcaps and found that their offspring's *Zugunruhe* was directed to a compass heading between the headings of their parents, but that the young birds retained the switch in direction of their Austrian parents. The two populations overlap in the wild but they maintain their separate identities because any naturally occurring hybrids would try to fly over the Alps, the Mediterranean Sea and the Sahara – a route they are most unlikely to survive.

Further evidence of the inheritance of migratory behaviour was demonstrated by breeding in captivity a population of French blackcaps which contained individuals that were either migratory or sedentary. Blackcaps that showed *Zugunruhe* were allowed to breed and in three generations all their offspring were migrants. At the same time, birds that did not show *Zugunruhe* produced entirely sedentary offspring in six generations. As well as this inheritance of a tendency to migrate, cross-breeding also showed that the amount of *Zugunruhe*, that is, the length of the journey, is inherited. The significance of these findings is that they show how the evolutionary process of natural selection can work on migratory behaviour, so that the birds adapt rapidly to changing conditions.

The blackcap is of particular interest to British bird-watchers because, over the last half-century, it has

increasingly been recorded through the winter, and has taken to visiting bird-tables where it feeds on bread and suet and learns to hang from peanut bags. Blackcaps that winter in Britain are not British natives but immigrants from northern and eastern Europe.

The blackcaps that winter in the British Isles have been shown to have *Zugunruhe* orientated slightly north of west, the course which will lead them across the North Sea. The advantage of this new migratory route has been demonstrated by keeping blackcaps under regimes of 'British' and 'Mediterranean' daylengths. The birds wintering in Britain have the benefit of longer days in spring and also have a much shorter distance to fly, so they will migrate back to Central Europe earlier and can establish territories before their Mediterranean relatives return.

These experiments show how blackcaps manage to make a single change in course but many migrants, such as the chaffinches that wend their way around the coast of the North Sea, have to follow devious courses. It is an even more difficult feat of compass work to fly on a Great Circle route (see pages 77–78). The advantage of navigating with reference to the Mercator projection is that the shortest route between two points lies on a single compass bearing, whereas flying on a Great Circle route requires the bird to steer on a continually changing bearing.

ABOVE *and* BELOW *German blackcaps migrate south-west and Austrian blackcaps south-east. Hybrids would take a middle course.*

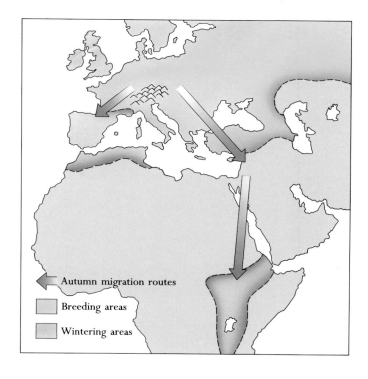

Autumn migration routes

Breeding areas

Wintering areas

PUTTING IT ALL TOGETHER

It is clear, then, that birds have an array of navigational devices. They use their eyesight, sometimes in ways that we have difficulty visualizing, they use a previously unknown magnetic sense, and they probably use their senses of smell and hearing. They use these senses to recognize landmarks and to fly on a compass course. Some of the information they need to reach a destination is inherited (see page 60) and some is learned. These elements need to be put together to produce a working system for guiding a bird around the world.

Although complicated by the fact that birds of different species will use their array of navigational aids in different ways, a general picture is emerging of how birds find their way. It appears to be important for a young bird to explore before it leaves on its first migration (see page 43), when it must be learning the sights, and possibly also the sounds, smells and magnetic fields, of its surroundings.

Evidence from the behaviour of young homing pigeons suggests that exploration of the home area teaches them landmarks which will later guide them on the last leg of a flight. Fitting pigeons with frosted contact lenses to blur their vision, and then tracking them with radio tags, is one of several experiments showing that birds can fly to within 0.5 to 5 kilometres ($\frac{1}{3}$ to 3 miles) of home without being able to see their surroundings, and then need landmarks for the final approach. It is possible that an array of landmarks forms a 'target' that will help the returning migrant locate its home if navigational errors cause it to fly back wide of the mark.

When the journey starts, the bird sets out on a pre-ordained course, perhaps modified by experience from previous years. To translate the directions coded in its genes into a heading, it needs to calibrate its compasses to show a co-ordinated sense of direction. The magnetic compass is the basic and most important of the set and is used to calibrate the bird's sun and star compasses, although they are used most in day-to-day orientation.

In at least some species, magnetic orientation develops shortly after hatching. Young racing pigeons can find their way home before they learn to use a sun compass at about 12 weeks old. During this learning process, the young pigeons use their magnetic compass as a reference while they watch the movement of the sun through the day. Their moon compass is also set against the magnetic compass, and some species (including white-throated sparrow, European robin and garden warbler) set their star compasses in the same way.

However, experiments with indigo buntings demonstrate that they rely not on a magnetic 'master compass' but on the movements of constellations across the sky to set up a star compass. Young indigo buntings were kept in a planetarium

The white-crowned sparrow is an American species that has been the subject of many experiments on navigation. It has been shown that it uses a star compass that is calibrated against a magnetic compass.

(where the stars are mimicked by the projection of dots of light on a domed ceiling), with the Pole Star in the centre. The young buntings learned to recognize the Pole Star's axis by the rotation of the other stars. If there was no rotation, they failed to find the axis, and if the rotation was changed to centre around another star, they adopted this as their 'Pole Star'. Once they had fixed an axis, they no longer needed to sense the rotation and could find north instantaneously.

In practice, the magnetic compass is secondary to sun and star compasses for flying on a heading. Perhaps the latter provide a more reliable reference than the magnetic compass, which can be upset (see page 66). It also appears to take time to take a bearing with the magnetic compass, whereas celestial compasses can be read instantaneously. The sun compass may be better than the star compass, at least for some species. Savannah sparrows orientate better if they have seen the sun set, when polarization in the evening sky is very strong, so they may be orientating first by the sun and then flying through the night using their star compass. The use of the two compasses in this way could be the explanation for nocturnal migrants taking off about half-an-hour after sunset.

A possible role for the magnetic compass may be to calibrate the celestial compasses at intervals so that the birds can cope with the changing positions of sun, moon and stars as their latitude changes during the flight. It is interesting that European robins, which are short-distance migrants, recalibrate their star compass only every few nights while indigo buntings, whitethroats and garden warblers that cross the equator and will see significant changes in the night sky, including the disappearance of the Pole Star, check with their magnetic compasses every night.

Celestial compasses also seem to be more accurate for keeping on course and the magnetic compass is retained as a back-up when the sky is blotted out by clouds or fog. Its great advantage is that it cannot be masked in nature. Migrants are often reluctant to take off when the sky is obscured but, if clouds or fog form while they are in flight, they keep going with the aid of their magnetic compasses.

Whichever compass is used, keeping on course is not enough to guarantee reaching the destination. As with a boat or a plane, a bird is at the mercy of the wind, which will drift it off course (see page 106). Its course over the ground in the direction of its destination is known as the *track*, and the wind pushes it off this course, so it must compensate by steering in another direction called the *heading*. The angle between track and heading is called the *drift*. You can sometimes see small aircraft or large birds flying crabwise as they compensate for a stiff side wind, heading into it to maintain their track.

Landmarks give birds reference points which enable them to adjust their headings, as is probably the case with the radio-tagged Canada geese described on page 17. This is another reason why migrants are reluctant to fly in overcast or foggy weather. Compensation for wind drift has been studied in migrating hawks, which are large enough and often sufficiently abundant to be tracked easily. To some extent, birds allow

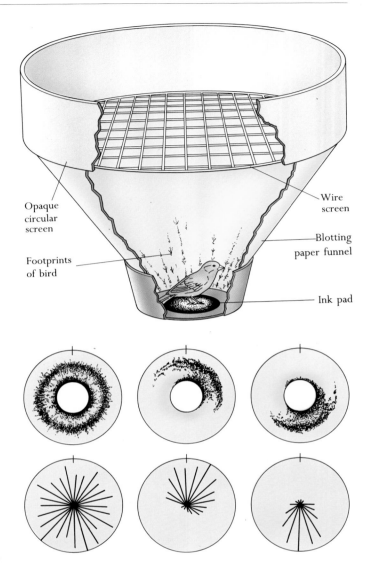

ABOVE *An indigo bunting in an 'Emlen cage'. The direction of its activity, recorded by inky footprints (or by scratches on typewriter correction paper), shows how it orientates relative to star patterns.*

themselves to drift with the wind. Provided that it is blowing roughly in the right direction, it will carry them quickly and economically to their destination. The birds will eventually have to change course to compensate for the drift off-course, and they may be helped by a change in wind direction. Broad-winged and other hawks migrating from Canada through the United States meet first westerly winds, then easterlies as they reach the southern states, so they drift first to the east and then to the west.

Compensation for drifting in the wind is more difficult at sea but some observations show that birds such as cranes (see pages 96–97) set their heading by the waves. Although the waves are moving in the same direction as the wind, their movement is much slower so the birds' rate of drift downwind will be much less than it would otherwise have been.

MAKING A SUCCESS OF FAILURE

Every year, salvos of birds launch themselves from their breeding grounds. Many hit the target, reaching their winter homes and then returning to nest the following spring, but many others miss the mark and die. Apart from those that starve or are eaten by predators or meet with an accidental death, many millions get lost because they are sent off-course by adverse weather or faulty navigation. Yet these failures are the raw material of evolution. They are being weeded out of the population by natural selection ('the survival of the fittest'), yet they provide the variability with which natural selection also allows animals to overcome potentially disastrous changes in the environment.

If a wintering site or a staging post disappears, there will be massive mortality, but the population may be saved by birds that have gone elsewhere and survived to colonize a new site. This is also the way that a species extends its range. The lapwings that are occasionally blown across the Atlantic perish in the severe winter weather of eastern North America, but fieldfares picked up by south-easterly gales over the North Sea in January 1937 successfully settled in southwest Greenland, and a small nesting colony survives there to this day.

Lapwings fleeing from cold winters in Europe are occasionally blown off-course and swept across the Atlantic to North America. Although they have not been able to survive to breed there, this is the way some species have extended their ranges.

CHAPTER 4

FLIGHT PLANS

A long journey must be planned. This does not, of course, mean that birds consciously plan their flights like an aircraft pilot, but that the plan for the journey has been established over the course of evolution and modified by the experience of previous flights. To arrive safely after a long flight is a marvellous feat of logistics. It requires more than setting off with a full load of fuel and the ability to navigate to the destination. These and other factors must be integrated to give the bird the best chance of accomplishing its aim, but luck still plays an important part.

These Arctic terns in the far north, in Alaska, have yet to embark on a record-breaking journey that will take them to the opposite end of the world and back.

PLANNING THE JOURNEY

One of the saddest sights for a bird-watcher is a migrant bird struggling to keep airborne over the sea but slowly losing height until it hits the water and drowns. This may be far out at sea when the bird is desperately trying to reach the deck of a ship but cannot catch up. It sinks lower and lower, dropping slowly behind, until swamped by a wave. Even more poignant is to see a bird hit the water within sight of the shore and journey's end.

These casualties are tiny tragedies which become trivial only when compared with the millions of birds that make the passage safely. They have, for some reason, failed to balance the delicate equations needed to make a safe journey and fatally run out of fuel. They may have started this leg of the journey with too little fuel or they may have met contrary winds. Maybe they should not have been there in the first place; they could have become lost.

It is not easy to piece together how a bird organizes its migratory flights. Bird-ringing (bird-banding in North America) gives only momentary glimpses, radar gives a wider picture, and radio tags come nearest to perfection in tracking the movements of individuals, but studying the broad strategies adopted by migrants helps to explain how small birds make transcontinental journeys.

BELOW *A long-eared owl rests on a North Sea beach. A stiff headwind might have prevented it from making landfall.*

THE GREAT CIRCLE ROUTE

Ringing recoveries and other observations provide evidence that wheatears follow Great Circle routes – the shortest routes across the globe – but make deviations to take advantage of prevailing winds.

The maps in an atlas can be misleading. Because the world is a sphere, areas have to be distorted when drawn on flat paper. Maps of small areas are almost the same as they would be on a globe but there are significant distortions on a continental scale. The cartographer tries to overcome this problem by using projections that involve attempts to draw the Earth's parallel lines of longitude and latitude on a flat surface. In Mercator's projection, which is traditionally used for nautical charts, lines of longitude (meridians) cross lines of latitude at right angles to make a rectangular grid, rather than curving to meet at the Poles as they do on a globe. The result is a map distorted near the polar regions, with Greenland appearing as large as the United States.

The shortest distance between two points on a Mercator projection is called the *rhumbline* but the shortest distance between the same two points on the globe is the *Great Circle route*. (A Great Circle is a line which divides the globe into two equal halves.) This can be seen on maps drawn on any gnomonic projection (based on a perspective from the centre of the Earth). When transcontinental airline flights were introduced, the old routes involving refuelling stops were abandoned in favour of shorter, non-stop Great Circle routes.

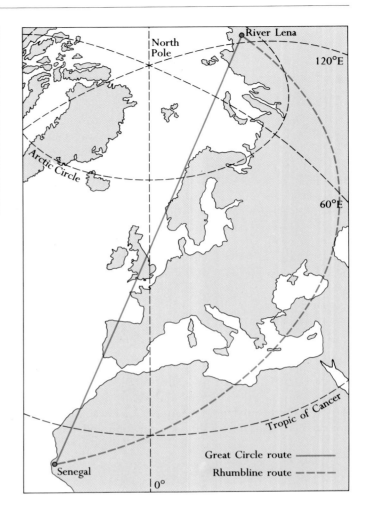

ABOVE *On a Mercator projection, the shortest route for a ruff flying between the River Lena, in eastern Siberia, and Senegal, in West Africa, appears to be the rhumbline. The Great Circle route looks far longer.* RIGHT *Replotted on a gnomonic projection (see text, page 77), the Great Circle route is seen to be the direct route.*

The Greenland ice-cap has become a familiar sight to passengers travelling between Europe and the Far East because it lies on the Great Circle linking these distant destinations. The advantage of Mercator's projection is that a straight-line route on the chart follows a fixed compass course. The aeroplane or bird on a Great Circle route must, by contrast, continually change its compass headings, but the savings in both time and fuel make it worthwhile.

Plotting bird migrations that follow Great Circle routes shows large deviations from what might be expected when looking at a conventional atlas. The shortest route on the Mercator map for a wheatear to fly from Baffin Island (northeastern Canada) eastwards, or from Alaska westwards, to central Africa, is not the shortest route on a globe. The eastward route on the globe does not take a long diagonal path across the North Atlantic but arcs over Greenland and sweeps down the west coast of Europe. Similarly, the westward route crosses central Asia, swinging north of the Himalayas and even entering western Europe.

The extent to which birds actually use Great Circle routes is still unknown. Ruffs nesting in eastern Siberia migrate to join others from Europe on wintering grounds around Lake Chad and the floodplains of the Senegal and Niger rivers in west Africa. On Mercator's projection, the direct, rhumbline, route is through central Asia and the Middle East, but the Great Circle route, which looks nonsensical on the map, takes the ruffs over the Arctic Ocean and down through Scandinavia and western Europe, and cuts nearly 20 per cent off the journey. So the ruffs which have been ringed in western Europe and recovered near the Lena river in eastern Siberia are not strays blown off-course but birds on their regular route to their breeding grounds. However, it has yet to be proved that the ruffs do follow the Great Circle route in its entirety. They may turn along the Siberian mainland to take advantage of refuelling stops, but this will be impossible to confirm until the birds can be tracked with radio tags.

Great Circle route ———
Rhumbline route -----

BELOW *According to the present state of our knowledge, the ruff does, indeed, migrate on the Great Circle route.*

DIFFICULT CROSSINGS

Many long-distance migrants are faced with major barriers to their flight. Asian birds pass to the west or east of the great peaks of the Himalayas (although some, such as the bar-headed goose, fly over the top), and American birds cross the Gulf of Mexico as an alternative to flying down the isthmus of Central America. At one time it was assumed that the huge ice-cap that almost covers Greenland was a barrier to bird migration, so that birds breeding on the island's west coast wintered in America and those breeding on the east coast headed for Europe. The Greenland ice-cap extends over 1.7 million square kilometres (656,420 square miles) and is a desert more inhospitable than the Sahara. However, observations made around the ice-free fringes of Greenland suggest that, despite its size and inhospitality, the ice-cap is not the barrier to birds it was once thought to be, and in 1936 the Danish explorer Eigil Knuth observed flocks of geese flying over as he skied across the ice-cap.

It is now known that waders and songbirds, as well as geese, nesting in north-west Greenland and neighbouring parts of Arctic Canada migrate on a Great Circle route across Greenland to Europe and Africa. Many of them stop to feed in Iceland on their return in spring but they do not continue on the Great Circle route up to the Arctic. They head first for the Ammassalik region of south-east Greenland and then turn northwards, probably along a rhumbline route, to their breeding grounds. This is 10 per cent longer than the Great Circle route, but it avoids the widest and highest part of the ice-cap – 1,500 kilometres (930 miles) wide and 2,500 metres (8,200 feet) high – where the winds are often contrary.

Although ice-caps, stretches of ocean, mountain ranges and deserts are not insuperable obstacles to bird migration, routes may be chosen to minimize their hazards. Migrants are funnelled across straits, round headlands, through mountain passes and along river valleys; in the process, they provide bird-watchers with fine spectacles of massed bird movements. Narrow sea crossings are especially famous in this respect. European birds form a concentrated passage across narrows such as that at Falsterbo, where they leave Sweden to cross the Baltic, or the straits of Gibraltar and the Bosporus, where birds leaving Europe avoid a long crossing of the Mediterranean. In North America there are similar funnels at Cape May, where migrants coming down the eastern coast leap across the mouth of Delaware Bay, and at Whitefish Point, where northbound birds cross Lake Superior and continue into Ontario.

Red-backed hawks and other soaring migrants avoid long water crossings and large numbers stream across straits.

THE GREATEST BARRIER

Of all the land and water crossings, the greatest natural barrier to bird migration is the Sahara Desert – up to 1,600 kilometres (1,000 miles) of barren waste with few places offering food and shelter. This greatest of all the world's deserts is an immense obstacle facing the migrants passing between the landmass of Eurasia and tropical Africa. It has been estimated that over 5,000 million birds must overcome this barrier every autumn.

Some of these birds must first cross the Mediterranean Sea; Asian birds fly over the Central Asian and Arabian deserts before reaching the vast expanse of the Sahara. The flight to the tropics, via the Mediterranean and Sahara, has long been a subject of interest because of its importance to European birds – and to ornithologists. As well as the physiological problems of long-distance flight, the birds face the complications of intense solar radiation and drought.

From what is known of the fuel consumption of small birds, their speed of flight and their ability to put on weight before migration (see pages 18–19), it is theoretically possible for some species to take off from southern, or even western, Europe and fly non-stop to a destination south of the Sahara. To achieve this, small birds must fly for 40–60 hours non-stop, having doubled their weight with fat to fuel the journey. There is still some argument about whether a single non-stop flight across the desert is feasible, however, because there are unknown factors in calculating the rate of fuel consumption and measurements of the flight speeds of migrants are limited. Some experts believe that assistance from tail winds is essential for a non-stop crossing. However, it is now known that birds fly more efficiently than was once believed (see box on page 87) and this is supported by evidence from radar observations that warblers fly 10 kilometres (6 miles) per hour or more faster than theoretical calculations had predicted. If this is so, a non-stop flight *is* feasible if conditions are favourable.

Prevailing winds blow mainly in the migrants' favour, especially for the autumn flight. The birds would have to wait for a north-westerly, which is not uncommon, before setting off across the Mediterranean, and they would then be carried by steady north and north-easterly trade winds of 20–40 kilometres (12$\frac{1}{2}$–25 miles) per hour over the desert. The return journey in spring is more difficult because the trade winds now hinder the birds. They have to wait for one of the moments when the wind swings round and then make a dash across the desert. Once they have reached the North African coast they can recuperate and make use of the flush of spring vegetation to build up fat reserves before crossing the Mediterranean.

Anyone who has travelled across the Sahara in spring or autumn will be struck by the number of birds to be seen on the ground. They are not only gathered at oases and in wadis, where there is a scattering of plants, but also along the desert

ABOVE *A wheatear searches for scarce insects on the desert sand, while waiting for cool air after nightfall before continuing its journey into tropical Africa.*

RIGHT *Yellow wagtails find a good roost in an Egyptian mangrove swamp where there is shelter and plenty of food.*

roads. There are birds, frequently warblers, wagtails or wheatears, hiding in the shelter of boulders or bushes, behind the petrol drums that mark the route, in the burnt-out wrecks of vehicles and even in the shadows cast by empty drinks cans and other litter. Ornithologists once believed that these birds were stragglers which had either set out underweight or had met head winds and run out of fuel, but there is an alternative explanation.

The notion that these birds were failures had been supported by the fact that those that had been caught were underweight, and the assumption was that they had been forced to stop because they had run out of fuel. Rather, it transpires that these lean birds are feeding actively by day and so fly into the ornithologists' nets, while the fat birds which have also stopped *en route* are not caught because they are not interested in feeding. They are quietly waiting in shelter until they restart their migration at nightfall.

The new conclusion is that trans-Saharan migrants do not land necessarily because they have run out of fuel. They fly at altitudes of perhaps 1,000 to 3,000 metres (3,300 to 9,900 feet), where cooler air reduces the stress of flight and they are likely to encounter tail winds that blow them along. If the winds change, the birds lose height to make better headway but then they face overheating and excessive water loss when the sun rises. Their best course of action in these circumstances is to land and find some shade until the evening. Some do run low on fuel and these are the birds found feeding at oases and taking the opportunity to top up their reserves.

ARCTIC TERNS: CHAMPION MIGRANTS

The Arctic tern is famous for making the longest migratory journey of any bird, but there are many gaps in our knowledge of its migration routes and the strategies it employs to achieve its remarkable feat. Most Arctic terns nest around the fringes of the Arctic Ocean. There are Arctic terns at the tip of Greenland, at Cape Morris Jesup, the most northerly land in the world. After breeding, the terns migrate to the southern summer of the Antarctic. Here, research vessels report them roosting in flocks on ice floes and fishing for krill (the little crustaceans that occur in huge numbers and support the populations of Antarctic whales, seals and seabirds) in pools of open water at the edge of the pack-ice.

From the midnight sun of the Arctic nesting colonies to the midnight sun of the Antarctic 'wintering' grounds is a distance of some 20,000 kilometres (12,400 miles). But the terns do not simply head south on a straight line and return the way they came. Their migration routes are not fully known and are based on sightings of birds seen on passage and on recoveries of ringed birds, but they are clearly complex. The main routes hug coastlines but some terns fly overland and others divert across oceans. Terns from the Baltic cross the Scandinavian landmass, and there is a recovery of an Arctic tern in the Colombian Andes at an altitude of 2,000 metres (6,600 feet).

Although Arctic terns are at home at sea, where they can alight to rest or stop to feed, they do not migrate leisurely and feed as they travel. Observations suggest that they travel like landbirds, making long flights between areas of sea where they stop to feed and fatten themselves. Their style of flight on migration is a striking contrast to the lazy, slow-flapping flight seen when they are foraging. Small flocks travel in V or echelon formations and fly, looking rather like waders, with shallow wingbeats and their wings flexed to reduce the span – a style of flight that is suited to rapid travel.

The first destination for Arctic terns from northern Eurasia is the sea off Norway and the British Isles, where a peak in the production of animal plankton at the end of the summer provides the birds with the opportunity to fatten up. They are joined by terns from the American Arctic which cross the North Atlantic after fattening up in the Davis Strait, west of Greenland. Together they head down the eastern seaboard of the Atlantic and pause again to feed off the west coast of Africa before making another long flight to the region of the Benguela Current off South Africa, where they feed again for the last time until they reach the plankton-rich Antarctic waters.

The terns' movements in the Antarctic are based on the prevailing winds. When they reach the edge of the pack-ice they drift eastwards, and those that came down the coasts of South America follow those that arrived via Africa. As the pack-ice recedes through the southern summer, the terns are brought into a zone of east winds and are blown back to the west. In March, they start their return flight, having moulted and put on weight, and are back in the Arctic by May or June, according to latitude.

LEFT *An Arctic tern reaches the Antarctic pack-ice, having migrated from one frozen ocean to another.*
BELOW *The terns' main migration routes take advantage of prevailing winds and good feeding grounds.*

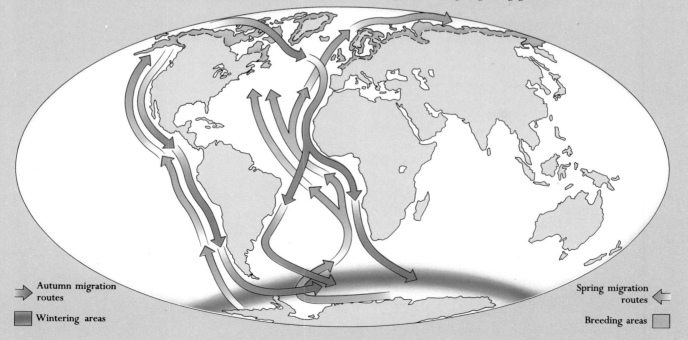

→ Autumn migration routes

▬ Wintering areas

Spring migration routes ←

Breeding areas ▢

SPEED OF FLIGHT

The main concern for a migrating bird, as with any traveller, must be to reach its destination before it runs out of fuel. It must carry more than enough to make a straight flight from point of departure to destination, because it must have a reserve in case it is held up or blown off-course. Assuming that it has fattened up sufficiently before departure, as described on pages 44–45, the migrant must use its fuel economically. Even when it has reached its destination before fuel runs out, it often needs a reserve to keep it going until it can continue feeding.

There is a link between speed and fuel consumption which is not immediately obvious but which is worth trying to understand because it is important for an appreciation of the strategy of bird flight. For an animal that runs or swims, higher speeds mean higher fuel consumption. The situation is complicated in flight, however, because energy is used to keep a bird airborne as well as to propel it forward. Part of the *lift* generated by the wings is caused by the airflow over their surfaces as the bird moves through the air. The faster the bird flies, the more lift its forward motion generates and it needs less power to support itself in the air by flapping. At the same time, *drag* (the resistance to its passage through the air) increases with speed and the bird has to flap harder to overcome it. The net result is that flying both very slowly and very fast is strenuous, but there is an economical range of speeds in between when the bird is obtaining plenty of lift from the airflow over its wings and drag is still fairly small.

The economical speed of flight varies between species, depending on size and wing shape among other factors. For each species there is a speed where power requirement is least. This is the *minimum power speed*. Flying at this speed allows the bird to remain airborne as long as possible on a *unit of fuel*. It is the speed of a bird that is not in a hurry but needs endurance. This might seem to be the requirement for a migratory flight, but to cover a long distance a bird needs to reduce the energy consumed per *unit of distance* and so fly as far as possible, rather than for as long a time as possible. By flying a little faster than the minimum power speed, incurring a little more drag and spending more energy, the bird can travel further on a unit of fuel. This slightly faster speed is the *maximum range speed*.

The relationship between power and speed is explained more fully in the box on page 87. It is based mainly on theoretical calculations but there is growing evidence that migrating birds do behave approximately as predicted. The figures at the top of page 87 compare calculated maximum range speeds with speeds observed by radar.

Geese are fast but not economical fliers unless they can find tail winds. These snow geese are coming in to land in front of Mount Shasta, California, after a long flight from the edge of the Arctic Ocean involving several stops on the way.

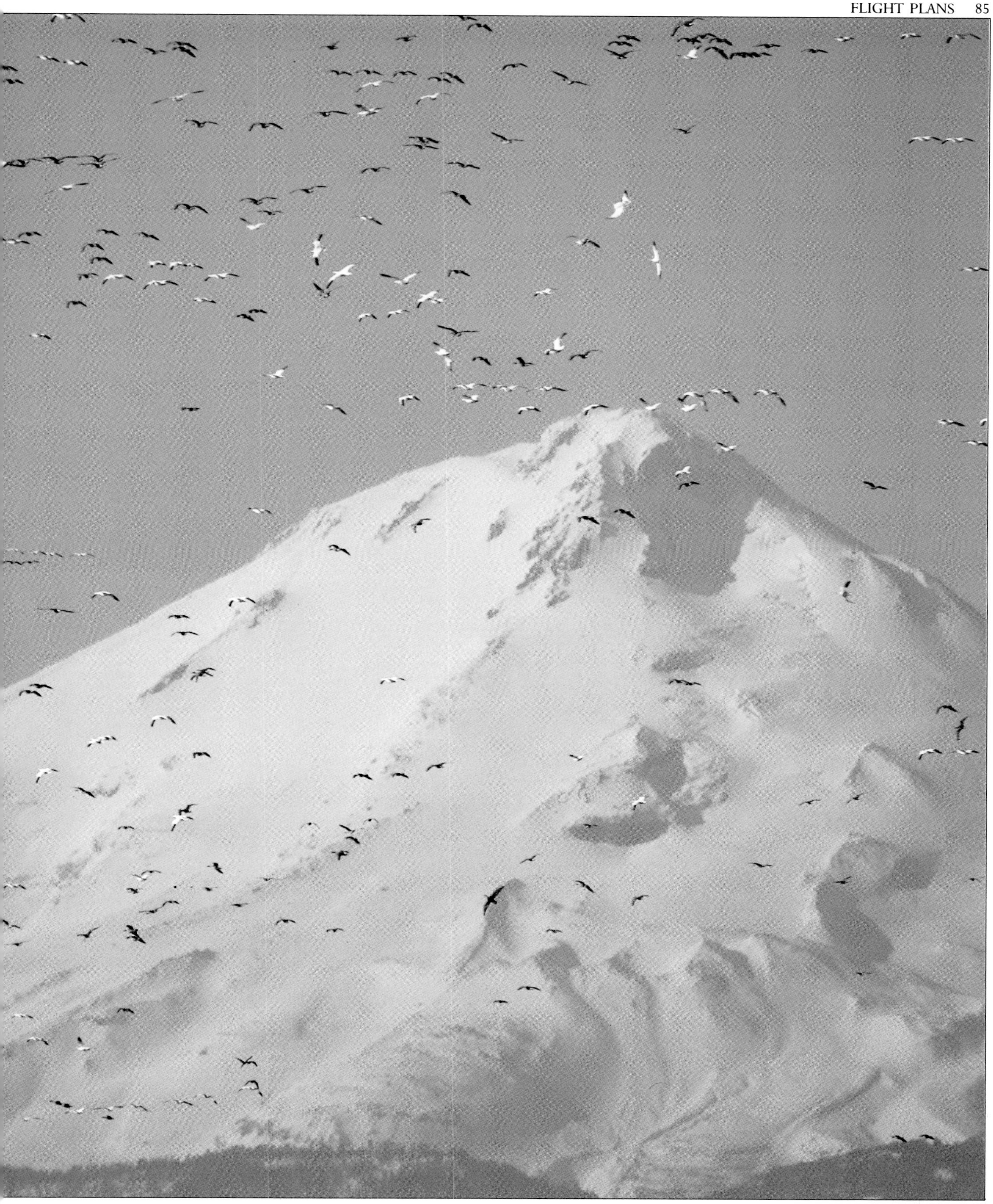

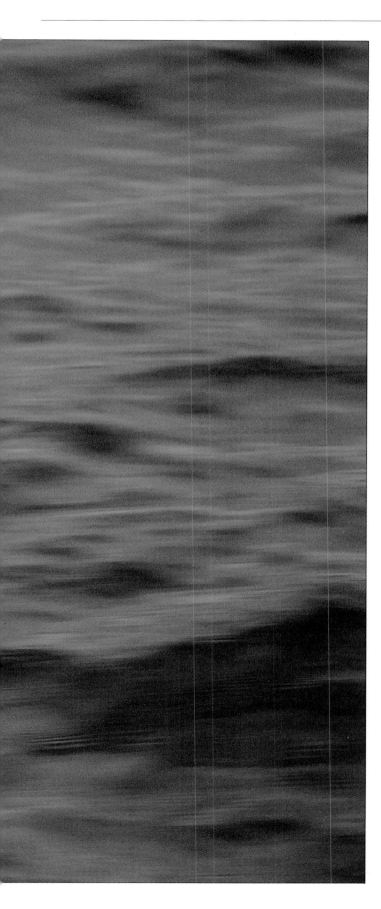

	Calculated maximum range speed km/hr	Observed speed km/hr
Common Crane	70	67
Eider	73	74
Swift	34	40
Chaffinch	34	39

These figures show that birds are, as predicted by theory, adjusting their flying speeds to cover as much ground as possible. This strategy should change if the birds become lost or are blown out to sea and cannot land. They must now slow

This Audubon's shearwater has shorter, less slender wings than most of its relatives, and does not make such long ocean journeys.

THE DYNAMICS OF BIRD FLIGHT

The theory of the basic relationship between power and speed in bird flight is shown by this power-speed curve. Like an aircraft, a flying bird has a higher rate of fuel consumption the faster it goes. Most of the increase is due to the increasing drag on its body from the slipstream. When it is flying very slowly, there is very little drag (or none if it is hovering), but it has to spend a large amount of energy flapping hard because it is getting very little lift from air flowing over its wings. The result is the U-shaped curve which shows that there is a speed, *the minimum power speed*, at which energy expenditure is minimal.

Very slow flight may be used when the bird is gathering food, and fast flight when it is escaping from predators or rivals. For sustained flight, it is economical to fly in the narrow range of speeds around the minimum power output. The exact range of speeds and power outputs depends upon the size of the bird and the shape of its wings.

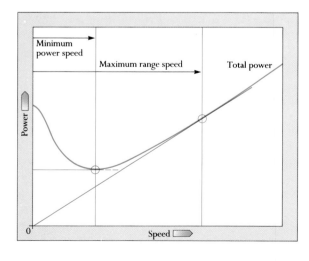

Migrating ospreys often visit rivers and lakes, where they spend a few days fishing before moving on again. Hovering to spot fish is very strenuous but the birds conserve energy while travelling.

down to the minimum power speed to conserve fuel and stay in the air as long as possible until they can get their bearings or make a landfall. This is presumably how birds manage to survive when they are swept across the Atlantic.

The difference between the maximum range speed and minimum power speed is usually so small and measuring a bird's airspeed accurately is so difficult, that it is not easy to decide which strategy it is using. Some proof that birds choose the speed appropriate to the circumstances comes from experiments in which white-throated sparrows were tracked by radar after they had been released from cages carried aloft by balloons. Under a clear sky, the sparrows set off in their usual migratory direction at a speed of 33 kilometres ($20\frac{1}{2}$ miles) per hour, which is the predicted maximum range speed. If it was overcast, the birds were confused and circled or set off in the wrong direction, flying at only 24 kilometres (15 miles) per hour, or less. This is near their predicted minimum power speed, the optimum economic speed for staying airborne until they can find their way to safety.

The theory of bird flight also predicts that migrating birds should adjust their airspeed according to the strength and direction of the wind. If there is a tail wind to sweep it along, a bird should fly slowly and save energy. It is *travelling* over the ground faster but *flying* through the air more economically. In a head wind, travel becomes more expensive. The bird has to fly faster and harder to maintain its speed over the ground. To visualize these situations, imagine first running down a 'down' escalator (equivalent to flying downwind), and then trying to run up it (flying upwind). The experiments with white-throated sparrows confirmed that birds behave according to this prediction. When released into a tail wind, their airspeed was 25–30 kilometres ($15\frac{1}{2}$–$18\frac{1}{2}$ miles) per hour but when faced with a head wind, they flew at 30–40 kilometres ($18\frac{1}{2}$–25 miles) per hour.

FLYING HIGH

Bird-watchers on the ground have a biased view of migration. Even with binoculars they will spot only those birds flying near the ground. These may be birds coming to the end of a stage in their journey and wandering about in search of somewhere to land. Bird-watchers may also see migrants hugging the ground to gain respite from a head wind, such as the finches seen flying over Minsmere (see page 51).

Evidence that birds are capable of flying at considerable altitudes came initially from pilots seeing birds flying near them, or finding identifiable remains stuck on their aircraft after a collision, and from mountaineers who reported seeing or hearing birds flying overhead. The true scale of high-altitude bird flight became apparent when radar began to record the movements of migrants. It showed that birds, including songbirds and waders, regularly fly at altitudes of 1,000–2,000 metres (3,300–6,600 feet) and sometimes up to 6,000 metres (19,700 feet) or even higher.

BELOW *Flight is more efficient at altitude and the great heights that birds reach can be appreciated when they are seen flying over mountains, like this bald eagle.*

One obvious disadvantage of flying high is that the bird has to consume a considerable amount of energy in gaining altitude, but the effort is worthwhile for several reasons. Theoretical calculations based on the power-speed curve (see box, page 87) show that flying through the rarefied atmosphere is less demanding. As the density of the air decreases, so does the drag on the bird's body. On the other hand, the bird must flap harder to remain airborne. The net result, however, is that the energy required to fly a particular distance is reduced with increasing altitude.

Additional advantages of flying high are being able to avoid gusting winds eddying around hills and mountains or bubbling thermals (rising warm air currents) that may disrupt small birds. Also, the birds will remain well above fogbanks and sandstorms, and even above clouds, that would interfere with seeing the sun or stars. Finally, there is an advantage to be gained from the fact that the atmosphere cools with altitude. The flight muscles generate so much heat that the body has to be cooled by evaporation of water from the respiratory system. This could lead to serious dehydration on a long flight, if it were not countered by the cooling effect of the air.

ECONOMY TRAVEL

A 'fair wind' may be as important to long-distance bird migrants as it was to human voyagers before the invention of the steam engine. Theoretical predictions of the distance birds can fly with a full payload of fat have led some ornithologists to the conclusion that the longest non-stop flights, such as those crossing the Sahara, may be possible only with the help of tail winds. Detailed studies of the bar-tailed godwit, which winters in West Africa, show that it could not carry enough fuel for its long, non-stop flight to the Dutch coast without the assistance of tail winds. The calculations of flight endurance that demonstrate the need for tail winds are reinforced by observations that many migrants wait for such winds before setting out (see page 54), and that they become grounded when winds turn against them.

There is evidence that birds not only choose to fly when the wind is in the right direction but also search the airspace to find the altitude at which the wind is most favourable. How they find the best height is not known, but the value of finding it is shown by the flock of swans (presumed to be whooper swans) tracked by radar down the coast of western Scotland and into Ireland at an altitude of 8,200 metres (26,900 feet). They were being blown along in a 180-kilometre (112-mile)

per hour airstream and had probably travelled from Iceland in only seven hours.

While it is a common observation that birds prefer to set off in a tail wind and are swept towards their goal with the least effort, there are other strategies for economical travel which involve adopting energy-saving techniques of flight. Choosing to fly at the most economical speed has already been discussed on page 84. Evidence that birds actually use this option is still weak but the facts of economic styles of flight are easy to see in the everyday activities of birds. However, the saving of energy has to be balanced against the reduction in speed that results from economic flight.

Savings in fuel consumption can be made with the correct wing design. A large wing gives a low *wing-loading* (the ratio of the wing's area to its weight) and makes for buoyant, less energetic flight. Its efficiency is increased if the wing also has a

BELOW *A weather map helps to predict bird migration. When possible, birds choose tail winds, so they use different sectors of a frontal system for spring and autumn migration. They must, however, avoid the worst weather in case they become diverted, injured or even killed.*

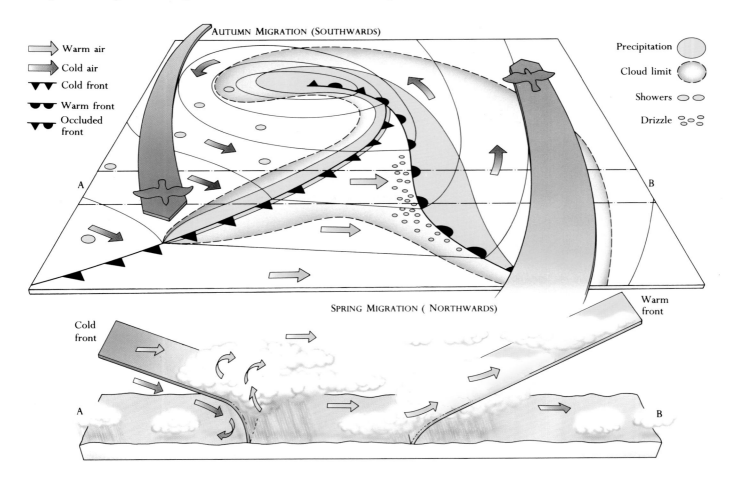

Warm air
Cold air
Cold front
Warm front
Occluded front

AUTUMN MIGRATION (SOUTHWARDS)

Precipitation
Cloud limit
Showers
Drizzle

A   B

SPRING MIGRATION (NORTHWARDS)

Warm front

Cold front

A   B

ABOVE *Long, slender wings are a feature of highly aerial birds. This royal albatross ranges over huge areas of ocean.*

LEFT *The short, rounded wings of Cetti's warbler are suited to manoeuvring rather than sustained flight; unlike long-winged relatives, this species never migrates.*

high *aspect ratio* (length/width) because the relatively narrow wing reduces drag. Drag is reduced even more if the wingtip tapers to a point.

Long, pointed wings are characteristic of such birds as swallows, swifts, terns, albatrosses, shearwaters and bee-eaters which spend much of their life in the air and require effortless flight. These birds hunt for food while airborne and they are long-distance migrants. High-aspect-ratio wings are also a feature of migratory waders and ducks that feed on the

ground. Swans and geese are too heavy for high-aspect-ratio wings. A swan with albatross-style wings long enough to support its heavy body would be dangerously weak and too clumsy for taking off and landing.

Songbirds have low-aspect-ratio wings but their large wing area allows them to carry considerable weights of fat without the wing-loading becoming uneconomically high. As a rule, they feed among vegetation, where a low-aspect-ratio wing is better for manoeuvring through foliage in short bursts of flapping flight. A compromise is reached in migratory species of songbirds which have relatively long and pointed outer portions of their wings (formed by longer primary flight feathers). Thus, in Europe, migratory sedge and willow warblers have longer, more pointed wings than sedentary Dartford and Cetti's warblers, while the medium-range chiff-chaff has wings of intermediate shape. Similarly, in the New World, the migratory Blackburnian warbler and American redstart have relatively longer wings than the sedentary rus-set-crowned warbler and yellow-crowned redstart. There can even be a difference between populations of partial migrant species. Blackcaps that migrate from northern Europe to Africa have longer, more pointed wings than resident black-caps in southern Europe. There is a similar trend in the New World yellow warbler.

SOARING AND OTHER TECHNIQUES

The most obvious form of energy-saving flight is the 'freewheeling' flight of birds that alternate gliding on outstretched wings with bouts of flapping. This is called *undulating flight*. The length of the gliding phase depends on wind speed and direction, and the bursts of flapping are needed to regain height and speed lost during the glide. Gliding is used to greater advantage in various forms of soaring by birds exploiting air currents to acquire extra lift. They are, in effect, extracting energy from the air and saving fuel.

One form of soaring, known as *slope soaring*, exploits updraughts formed when the wind rises over hills and ridges. Birds glide at speed along the line of the raised ground, buoyed up by the ascending air. The best exponents of this habit are hawks and the most famous place for watching migrants using slope soaring is Hawk Mountain, Pennsylvania, where thousands of hawks pass each day in the autumn. Hawk Mountain is at the southern end of the Kittatinny Ridge, which runs for several hundred kilometres from New Paltz, New York State, in a direction that is suitable for migration and forms a highway for the hawks.

Another form of soaring much used by hawks and other large birds such as storks, cranes, herons and pelicans, but also by smaller birds, is *thermal soaring*. The birds use the bubbles of warm air, called *thermals*, that rise through the atmosphere when the ground heats up. They circle within a thermal,

White pelicans soar overhead on their broad, outstretched wings. They use thermals to carry them aloft and then glide effortlessly for long distances, expending little energy in the process.

supported by the column of ascending air in the centre, and are carried aloft. Eventually the thermal weakens and the birds glide to the next one and repeat the sequence. It is something of a mystery how birds find thermals. Their position may be given away by the formation of cumulus clouds above them, but the birds must be very sensitive to air currents to find the precise place. They may help each other by travelling in flocks so that all can take advantage of a thermal that only one finds.

With the aid of thermals, these birds can fly all day with

ABOVE *Swainson's hawk is one of several species of hawks that migrate through America. By soaring in thermals and using economical flapping-and-gliding flight, it may not have to stop to feed once on its journey.*

hardly a wingbeat so there are substantial energy savings. Swainson's hawks can migrate from New Mexico to Argentina (approximately 8,000 kilometres or 5,000 miles) in two months, taking enough fat from their point of departure to sustain them throughout the journey without the need to feed on the way. (Some must fail *en route* because individuals have been found in Argentina so emaciated that they have been captured by hand.) The disadvantages of travelling by thermals are that they do not form in cloudy weather, so migrants that rely on them may be grounded for days, and they rarely form over the sea, so soaring birds cannot make lengthy sea crossings. This is why narrows such as the Strait of Gibraltar or the Dardanelles in Turkey are such good spots for watching migration. Spectacular numbers of soaring birds ascend in thermals on one side of the crossing point and glide across the water to pick up the thermals on the other shore.

Seafaring albatrosses, shearwaters and petrels, and other birds to a lesser extent, use a form of energy saving called *dynamic soaring*. They glide downwind, gathering speed, then turn sharply and climb into the breeze, using the airflow to give them good lift. Although they soon lose headway, they gain some extra lift because they meet progressively stronger winds as they climb through the first 15 metres (50 feet) above the sea surface. This allows the birds to regain their original height before turning downwind for another glide. These

Below A flock of Canada geese flies efficiently in V-formation. Except for the leader, each bird is assisted by the slipstream of the bird in front and so reduces the effort of flight.

gliding seabirds are also helped by a form of slope soaring in which lift is provided by the waves. Even in a flat calm, the movement of the swell causes an updraught that seabirds can use to extend their glides.

Undulating (flapping-and-gliding) flight is most used by larger birds, such as hawks and herons. It is uneconomic for small birds which use, instead, *bounding flight* in which they alternate between flapping and closing their wings. The result is an exaggerated bouncing which is very striking in flocks of finches. The theory, which is unproven, is that the drag on the wings of small, broad-winged birds is so great when they glide that they do better to fold their wings and drop, then climb again with a bout of flapping so that they follow a roller-coaster course through the air. It also seems that small birds cannot vary the power output of their flight muscles so it is more economical for them to fly with bursts of flapping at the most efficient, although expensive, wingbeat rate followed by economical 'freewheeling' with the wings folded.

A final labour-saving trick is to fly in a V or echelon formation. Calculations show that the upwash of air from a bird's wings has a lifting effect on its neighbours. The effect is greatest if the flock is flying in very close formation, but it can still be felt in the ragged formations typical of cranes, geese, swans, pelicans and other large birds. The energy savings may be small but they would become significant on long flights.

TRACKING MIGRATING CRANES

In 1978 the British bird flight expert Colin Pennycuick and the Swedish bird migration expert Thomas Alerstam used a light aircraft to follow cranes in spring as they migrated northwards through Sweden. The common crane uses thermal soaring on migration but not so exclusively as the rather similar white stork. The latter is grounded in weather that does not permit the formation of thermals. It travels only by day (but see page 57) and is forced to make short sea crossings. The more versatile crane has a wider choice in selecting how it will travel.

Some of the Swedish population of common cranes migrate in autumn to south-western Europe, and sometimes into North Africa, while the remainder join birds from Finland and Russia and follow a more easterly route to the Middle East and the Nile Valley. The cranes heading on the westerly route cross the Baltic Sea to the German island of Rügen (following the ferry route!) and then travel south through Germany and France. If the winds are favourable, they may reach France in one hop; otherwise they use traditional stopover sites.

Once the cranes have landed, they may stop for as long as there is plenty to eat, but they are very aware of the weather and set off again as soon as a favourable weather system arrives to carry them onwards. The destination for many of these birds is Les Landes, near Bordeaux, France. Formerly a wild region, it now grows crops of maize where the cranes feed in the fields and retire to roost safely within the perimeter of a high-security military area. In recent years more cranes have spent the winter here and only a few continue across the Pyrenees as winter advances.

Pennycuick and Alerstam followed the cranes when they returned to Sweden in April. The flocks gather at Rügen and set off across the 80-kilometre (50-mile) crossing of the Baltic Sea. Their first destination is Lake Hornborgasjön, in southern Sweden, about 400 kilometres (250 miles) to the north. If they set off from Rügen after daybreak, most of the cranes will arrive at Hornborgasjön by sunset.

The cranes wait for favourable weather, then flap steadily across the sea. On reaching land and the rising thermals they switch to soaring. Like a wave of invading bombers, the flocks were picked up on radar and the two ornithologists took off in their aircraft to intercept them. Careful manoeuvring was needed to keep the much faster plane in contact with the cranes and permit observations.

In fine weather, the cranes start looking for thermals and rapidly gain height. When they reach the top of one thermal – usually between 500 and 1,300 metres (1,640 and 4,260 feet), but sometimes over 2,000 metres (6,560 feet) – they set off in a straight line until they find another. The flock keeps together all the time, flying in a close V or echelon between the thermals but closing into a tight phalanx and spiralling together inside them. Between thermals, gliding is interspersed with flapping, which becomes more frequent

when the wind is less favourable. The cranes also have to flap almost continuously in bad weather or when the clouds formed by the thermals coalesce into a sheet and prevent further heating of the ground.

Radar observations show that, when crossing the sea, cranes fly at an airspeed averaging 67 kilometres (41½ miles) per hour, which fits in with theoretical calculations (see page 87) and, depending on the wind, they cross the Baltic Sea at a groundspeed of 56 to 102 kilometres (35–63 miles)

per hour. Similar speeds are recorded when the cranes are flying over land without the benefit of soaring. When they are using thermals their speed is much lower. They glide at about 70 kilometres (43 miles) per hour but, when soaring is taken into account, their overall airspeed drops to less than 50 kilometres (31 miles) per hour.

Cross winds slow down the cranes when they are over the sea because the birds are less capable of holding their course. When out of sight of land, they appear to use the waves as reference points to hold their heading. As the waves are moving downwind, the cranes slowly drift off-course, although not as much as if they were making no compensation for the wind. On reaching land, they can take bearings on topographical features and their deviation from the course for Lake Hornborgasjön is greatly reduced.

A flock of common cranes glides in neat formation while the birds search for a thermal to carry them aloft.

JOURNEY TIME

There is a significant difference between the average speed of a car when it is moving along the road and the overall average speed of a journey, because of delays and stops *en route*. Similarly, the time taken on a bird's migratory flight is not simply the distance divided by the mean flight speed. While some migrants set awesome records for speed and endurance, they are usually making long sea crossings where they have little option except to keep going at the maximum range speed. Other birds proceed by a series of leisurely stages and spend one or more days resting and feeding at stopping points. As a result, thrush-sized birds travel about 150–200 kilometres (90–125 miles) per day and warblers about half this distance, although there is considerable variation. Some individuals have been recorded, by lucky recaptures of ringed birds, as covering twice the average daily distance. Larger birds such as geese and swans have much slower journey times, despite faster flying speeds, because they spend more time feeding and fattening at stopping points.

The journey time for the migration takes into account these stops, as well as the time that the birds spend putting on weight before departure. This is why the actual time for migration is often measured in weeks when the birds' flying speed indicates that they could do it in days. Waders have faster journey speeds than songbirds because, although they put on weight more slowly, they fly faster. Typically, waders make fast long-distance flights between scattered estuaries and bays where they find mudflats for feeding. Songbirds move in shorter but more frequent steps and usually make long 'hops' only when faced with wide stretches of sea or desert.

A bird that takes off with a full load of fat is hampered by its weight. By the time it arrives at its destination, it will have used as much as one-third more energy than a bird that started with less fat and stopped to refuel on the way. Therefore, unless there are good reasons to the contrary, it is better to travel in stages. The problem with travelling light is that stopping places for migrant birds are often not so regular or reliable as motorway service stations or airports. A failure in food supplies at a staging place leaves birds in a difficult situation that could spell disaster.

BELOW *Turnstones and dunlins break their journey to feed at the tide's edge and refuel for the next stage of their migration.*

STAGING POSTS

The 100 million wildfowl that migrate from the north to the southern United States and Mexico each autumn break their journeys at lakes, ponds, rivers and marshes. These wetland habitats started to disappear in the 1930s on a large scale, mainly because of drought but increasingly through drainage for agricultural or building development. In 1934, the US Congress passed the Federal Migratory Bird Hunting Stamp Act, which specified that all goose and duck hunters are required to purchase a Duck Stamp, currently valued at $12.50. Ninety per cent of the revenue raised by the stamps is used to purchase and maintain refuges for wildfowl.

Staging posts for migrating birds must fulfil two functions. They must provide plentiful food and safe roosting when the birds are not feeding. Estuaries and bays with mudflats and sandy beaches are important staging posts for many species of waders which stop to feed on the rich populations of invertebrates living under the surface or washed in by the tide. They are of limited use as roosts because they are covered by

RIGHT *Adult bluethroats migrate faster than juveniles because they put on weight more rapidly at staging posts.*

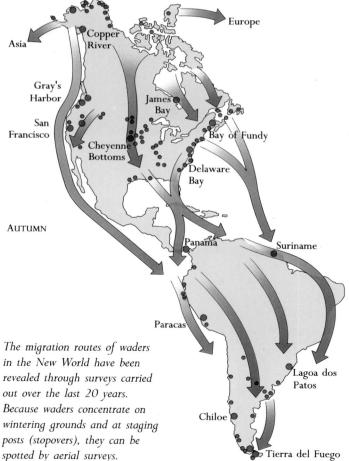

AUTUMN

The migration routes of waders in the New World have been revealed through surveys carried out over the last 20 years. Because waders concentrate on wintering grounds and at staging posts (stopovers), they can be spotted by aerial surveys.

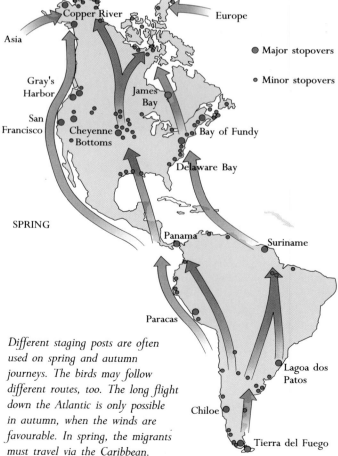

- ● Major stopovers
- • Minor stopovers

SPRING

Different staging posts are often used on spring and autumn journeys. The birds may follow different routes, too. The long flight down the Atlantic is only possible in autumn, when the winds are favourable. In spring, the migrants must travel via the Caribbean.

Above *The rufous hummingbird has to stop to feed frequently on its 3,500-kilometre (2,200-mile) journey from Alaska to Mexico.*

water twice a day. One of the most exciting yet predictable sights in bird-watching is to stand at the edge of the shore and watch dense flocks of thousands of waders flying inland as they are pushed off the mud by the rising tide.

While birds may also interrupt their migration because of unfavourable travelling conditions (see page 106), regular staging posts give them the chance to fatten up before the next stretch of their journey. The speed at which they can refuel will have a crucial impact on the overall journey time. The rate at which birds put on weight can be remarkable, with songbirds and waders regularly putting on 1 to 2 per cent of their lean weight each day. Gaining weight is dependent on the rate at which the birds feed and metabolize the food into stored fat.

The time spent at a staging post depends on how quickly a bird can refuel. Feeding may be limited to daylight hours or, for waders, to low tide. It can be hampered by bad weather, for instance through snow covering fallen seeds, and strong winds buffeting long-billed waders so they cannot probe efficiently for buried invertebrates. Even with a superabundance

of food, there is a limit to what a bird can eat at a sitting. Rufous hummingbirds stopping in California spend three-quarters of the day perching because of the time it takes to empty the crop of a feed of nectar.

When food supplies are limited, competition results. This was observed in individuals of the northern waterthrush, a species of wood warbler, that stopped in a grove of trees surrounding a pond on the Aransas River floodplain in Texas. The waterthrushes passed through in loose flocks but those that stayed often set up territories and became aggressive to each other. Repeated capture and weighing showed that only those with territories put on weight. On average, a waterthrush had to wait a couple of days to acquire a territory. It then put on weight at a rate of 7 per cent of its body weight per day and left about three days later. However, its departure was less dependent on its weight than on the arrival of suitable weather for its onward journey.

DELAWARE BAY

Most of the New World population of knots spend the 'winter' in Argentina. In spring they fly north over Brazil and Surinam and along the south-eastern coast of the United States. By late May, the knots have gathered in Delaware Bay, which is a major staging post. Aerial photographs of the flocks reveal numbers in excess of 100,000, which represents a third to a half of the entire New World population of the species. The knots are not the only waders to arrive. Large flocks of turnstones, sandpipers, plovers and many other birds form, too.

The attraction of Delaware Bay is that late May is the spawning time of the horseshoe crab, a 'living fossil' that has remained unchanged for 300 million years. On successive nights at full moon, when the tide is at its highest, the beach becomes alive with horseshoe crabs swarming along the water's edge for many kilometres to lay their tapioca-sized eggs.

One problem facing waders migrating northwards in spring is that the numbers of invertebrates living on the shore will have been reduced by birds and other animals that feed on them through the winter. The eggs of the horseshoe crabs are a renewed bonanza on which the waders can gorge to repletion. Each wader consumes 100–200,000 eggs to double its weight in two weeks, and then sets off for its nesting ground in Arctic Canada without the need for further stops, and still with enough stored energy to lay a clutch of eggs.

Knots settle in swarms at sites such as Delaware Bay that provide good feeding, and put on weight rapidly for the next stage of migration.

CHAPTER 5
HAZARDS AND DELAYS

When a bird sets out on migration it faces many hazards. Despite the care it takes in making preparations for the journey, and in timing its departure, and despite its navigational skills, the outcome of a migratory flight must still be a lottery. There is so much that can go wrong on the way and millions of birds perish every year. The greatest losses, however, are among young birds on their first journey and, spectacular as such losses may be in special circumstances, migration is the better option in the long run than remaining in one place.

Spotted redshanks fly over a sunlit estuary. They cannot expect such perfect conditions throughout their long migratory journey.

SALUTARY STORIES

In the spring of 1991, many British bird-watchers became worried because their favourite migrant birds were failing to return from their winter quarters in Africa. The most obvious losses were of swallows and house martins, which nest around people's homes, but there was also a dearth of warbler song in the woods and along the hedgerows. One of the problems was wet, stormy weather over the Mediterranean during April, at the time when millions of small birds are flying north on a broad front out of Africa.

Twenty years earlier, in September 1971, a family of three young white storks appeared in south-west England. They bore rings that could be read through telescopes to show that they came from Denmark. For some reason, the storks had set off in a westerly direction, crossing the North Sea and the width of England, instead of heading southwards to the species' wintering range in sub-Saharan Africa. One stork fell down a chimney. The two survivors eventually disappeared from Britain but one was picked up in a poor condition on Madeira about a week later and taken into care. It seems that it had reorientated and was heading in the right direction.

These two incidents, on large and small scales respectively, show that, for all its advantages, migration is a hazardous business. Despite the dangers, however, it must be easier than trying to survive through the winter on the breeding grounds; otherwise it would not be attempted. Migration evolved in situations where it is a better option than staying at home and it is not surprising to find birds ceasing to migrate when given the chance. A good example is the mourning dove of the United States. This species now survives winters in New England, at the northern end of its range, by taking advantage of spilt grain and the offerings at bird-feeders.

Calculating the extent of the many hazards that face migrating birds is an impossible task. Where birds, such as terns or sand martins, nest at traditional nesting sites, some idea of mortality on migration can be gained by ringing individuals before departure and recording the survivors that return to nest the following spring. This method does not allow for birds that died on the wintering grounds rather than on the journey, however.

Fewer sand martins live to return to their European breeding colonies if they meet frosts, storms or drought on their autumn migration to Africa. Yet it is better to chance these hazards than try to survive the northern winter.

THE SCALE OF DISASTER

Ornithologists at bird observatories on the Atlantic coast of North America estimate that several million small birds fly out to sea each autumn and are drowned. Some struggle back to the shore and are recorded along the south-eastern coast, while a few alight on ships and are carried to safety – unless the ship is going the wrong way. The liner *Mauretania* sailed from New York in October 1962 through strong west and north-west winds from the tail of Hurricane Daisy. When it was one day out from port, the ship was 'echoing with bird calls, and parts of the open deck space (were) almost inundated at times with small passerines.' There were at least 130 birds of 34 species, ranging from wrens to woodpeckers and a merlin. Despite food being put out by the stewards, most of the birds had died or disappeared by the time the ship reached the British Isles, but nine survived to fly ashore. The *Mauretania*'s stowaways represented a tiny fraction of the birds that must have been swept out to sea on that occasion.

Disaster also strikes when the weather holds up migrants. There was a celebrated incident in September 1931 when a cyclone over Poland caused a vicious cold spell, with heavy rain in southern Germany, Austria and Hungary. Swarms of swallows, house martins and sand martins massed in the shelter of buildings, where they became so torpid that they could be caught by hand. In Austria, the public were asked to collect the stricken birds and take them to Vienna Zoo, where 89,000 were fed on mealworms and transported by plane and train to Venice for release.

Another sudden catastrophe was recorded in March 1906, when a sudden heavy fall of snow in the neighbourhood of Worthington, Minnesota, hit migrating Lapland longspurs. In what became known locally as the 'great bird shower', an estimated one-and-a-half million individuals of this bunting species (known in Europe as the Lapland bunting) were killed in the course of a single night. They had become totally disorientated and were gathering around street lights and crashing into buildings. Not all died: a nightwatchman dug some out of the snow, dried their plumage and released them in the morning. Also, a Mr Dorbeck noticed lumps of snow on his roof and, as the morning sun began to melt them, the heads of birds appeared. These fortunate buntings eventually emerged, preened for a while and flew away.

A peregrine blown out to sea has found refuge on a ship. It has also found prey, so its chances of survival have improved.

FOUL WINDS

'And there went forth a wind from the Lord, and brought quails from the sea, and let them fall by the camp, as it were a day's journey on this side, and as it were a day's journey on the other side, round about the camp, and as it were two cubits high upon the face of the earth.' This verse from the Bible (Numbers 11:31, 32) is the first record of a drift of migrant birds, in this case one of such a size that it provided the wandering Children of Israel with a welcome change from their diet of manna. Quails migrate on a broad front across the Mediterranean in prodigious numbers, and 'falls' of flocks blown by the wind and settling over a broad area have occurred in modern times (see also page 110). Birds have difficulty in allowing for wind drift when crossing water and a long sea crossing, such as that undertaken by quails flying between Europe and Africa, gives ample opportunities for migrants to become displaced.

The problem for birds that embark on a long flight with the wind behind them is that they may suddenly find themselves in peril. The prevailing wind may blow steadily for days and encourage them to take off but it can change with hardly any warning. The birds may be able to compensate by heading into the wind, but too strong a wind will drift them off-course. Provided that they are still fat or can find food while they wait for a change in the wind, they will set off again, reorientating and heading for their destination.

If the wind is against migrating birds, it slows them down and uses up their energy reserves. This is when bird-watchers on the coast see migrants flying in low, hugging the waves where the wind will be slowed by friction. The land behind the shore becomes a refugee camp for exhausted birds, which swarm in the bushes and long grass and barely have the strength to move when flushed. If the birds are flying over land when met by a contrary wind, it is easy enough for them to halt at an impromptu staging point and perhaps have a chance to feed.

When, in October 1979, a flock of 200–300 chimney swifts descended on Islas del Cisne, two small islands 180 kilometres

ABOVE *The Ipswich sparrow, a subspecies of the savannah sparrow, normally winters along a narrow strip of the eastern US coastline, but one has been swept across the Atlantic to Britain.*

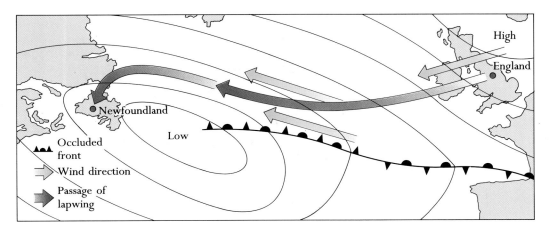

High
England
Newfoundland
Low
Occluded front
Wind direction
Passage of lapwing

LEFT *In 1927, a lapwing ringed in northern England was swept across the Atlantic to Newfoundland. The sequence of events began when the weather in England turned cold, and the bird flew west into Ireland in search of milder conditions. Then, a deep depression coming across the Atlantic created easterly gales which picked up the lapwing and carried it rapidly across the ocean to the New World.*

(112 miles) north of Honduras, they remained there until they died. The flock must have been blown over from their regular migration route down Central America and, with little insect food on the islands, the swifts went into a decline. After a week or so, all had died of starvation, although it was only five or six hours' flying time to the mainland. They had been

BELOW *A rare, secretive bird in Europe, this Baillon's crake was blown off-course and eventually turned up in an urban park in eastern England, to the surprise and delight of bird-watchers. As with other crakes, its usually skulking habits make it very difficult to study, and there are large gaps in our knowledge of its migration routes. Migrants have been seen in winter along the whole length of the Mediterranean basin, and probably the normal winter quarters of most European breeders are in sub-Saharan Africa.*

exhausted on their arrival and it seems that they were unwilling or unable to attempt the long, unknown sea crossing to reach a safe haven.

Contrary winds are non-selective and a hazard for any bird, but faulty navigation is most likely to cause juveniles to stray off-course. Any young bird that lacks the ability to reach its destination is unlikely to reach adulthood. Transatlantic strays brought across to Europe by the weather will find it virtually impossible to return to America in spring because the prevailing winds are against them. However, vagrants from Siberia that arrive in western Europe, although wind-assisted for part of the journey, appear to have come the wrong way because they made errors in navigation; it is believed that a few individuals of some species some may survive to return home (see page 111).

STRAY BIRDS ON THE SCILLY ISLES

On 20 October, 1986, a North American grey-cheeked thrush landed in the Scilly Isles, 45 kilometres (28 miles) off the south-west tip of England, after a 4,800-kilometre (3,000-mile) flight across the Atlantic. Its tragic ending made the newspapers: the next day, in front of about 200 bird-watchers, it was caught and killed by a cat. The thrush was doomed anyway because it had been heading for South America until it was swept across the Atlantic, and its chances of returning across the ocean were negligible.

During the 1960s, the bird observatory at St Agnes on the Scilly Isles was receiving occasional records of American species landing on the islands. The possibility was raised that the birds were being brought across the Atlantic on ships. It was well-known that liners like the *Mauretania* often carried avian stowaways (see page 105) and they sometimes steamed very close to the Scilly Isles. However, there was no particular correlation between the passage of the ships and the appearance of American birds on land. No doubt some birds do hitch a ride on ships for part of the way but the majority that reach the Scilly Isles and other parts of the British Isles have done so naturally. With the exception of seed-eating buntings and American sparrows of the family Emberizidae, which more often turn up in spring and probably with ship-assisted passage, most of these vagrants are young birds that have run into trouble on their first autumn migration and found themselves over the Atlantic Ocean.

The importance of the Scilly Isles as a haven for stray birds lies in their position off the Atlantic coast. The birds that make the safety of the islands must represent only a small minority of those flying over the Western Approaches and desperately seeking a haven. The islands form a target visible over a wide radius and birds are drawn from a huge area and concentrated in a small space that offers shelter.

The economy of the Scilly Isles has rested on catering for summer tourists and growing spring flowers. There was a dead period in late September and October after the tourists had departed but this has now been filled by bird-watchers arriving in hundreds, and even thousands, to add American, and sometimes Asian, species to their lists. After an initial period of strained relations between visitors and inhabitants, when fields were trampled and gates left open, the autumn surge of bird-watchers requiring accommodation and transport has extended the tourist season by two months. The misfortune of doomed American birds is a dividend both for keen European bird-watchers and the inhabitants of the Scilly Isles.

BELOW *Many American birds that have been swept across the Atlantic by autumnal storms make landfall on the Scilly Isles. They attract crowds of 'twitchers' — bird-watchers keen to add new species to their tally.*

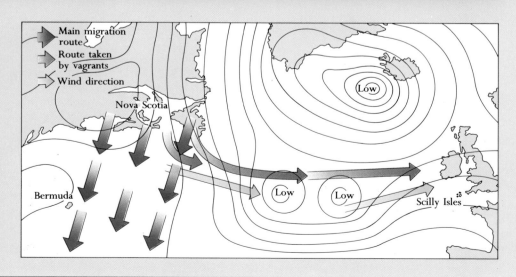

ABOVE *A grey-cheeked thrush eats a pokeberry in its native land. This is one of the many migratory American species which have been spotted as vagrants in Europe. They have little chance of returning home.*

RIGHT *Migrants that set out from the eastern seaboard of North America in autumn and head for South America run the risk of being caught up in westerly winds generated by depressions in the Atlantic and carried to Europe.*

FALLS

Martin Duncan was one of the British pioneers of migration study in the early years of the twentieth century who watched the nocturnal passage of birds in the beams of lighthouses. He has described the thrill of 'the 'endless procession of birds' and goes on to say: 'Hour after hour, literally hundreds of small birds, migrating warblers, wheatears, goldcrests, pipits, redstarts, whinchats and the like, arrive and pass on their way.' The drawback to such early studies was that observations at lighthouses show just abnormal movements. The main stream of migrants passes high overhead and the birds are attracted to the beams of a lighthouse only when visibility is poor.

The appearance of a mass of birds is known as a 'fall' or 'rush' and it can be extremely spectacular. There was an amazing fall of migrants on the south-east coast of England, mainly in Norfolk and Suffolk, on 3 September, 1965. The regular stream of birds that heads south-south-west from Scandinavia to Spain and Portugal and beyond encountered a zone of foul weather that disorientated, concentrated and finally forced them to ground in a belt of heavy rain. It was estimated that more than half-a-million birds came down on one 39-kilometre (24-mile) stretch of coast. About 80 species were involved, including many scarce ones, such as wrynecks, bluethroats and dotterels. The town of Lowestoft witnessed a cloud of birds passing over, with individuals dropping out until the streets and gardens were alive with birds. Many of these arrivals were exhausted and some even landed on people's shoulders! Others were washed up on the beach, presumably representing a minority of those that had perished in the sea. However, as soon as the weather improved, the majority of survivors were able to continue on their way.

Provided that they can land safely, refuel and reorientate, the birds in a fall will be safe, but man-made obstacles can prove a fatal hazard. Modern high-rise buildings have a similar effect to lighthouses, while TV transmitter masts act as passive 'nets' by intercepting birds with their guy wires. (On a single morning, 2,117 birds of 37 species were picked up under a mast in Eau Claire, Wisconsin.) Some lighthouses are notorious for the numbers of birds that are killed by dashing into the lantern and it was originally feared that the flares on oil-rigs would cause mass deaths as birds flew into the flames like moths into a candle. In the event, it has been found that the birds circle the flare without coming too close, and the danger is that they will become exhausted and fall into the sea.

BELOW *Like many birds, dotterels may become disorientated during migration. In certain conditions, they are forced down in large numbers. Obstacles such as wires can cause heavy mortality.*

GOING THE WRONG WAY

Sometimes errors of navigation cause birds to migrate in the wrong direction. For many years a black-browed albatross has been spending the summer among gannets nesting on the Shetland island of Unst. As it is a native of the southern oceans, something must have gone wrong with its navigation to have brought it so far from home. Although the albatross has built a nest, the chances of it finding a mate are almost zero (although there was a suggestion that one might be shipped up from the southern hemisphere).

The black-browed albatross is one of the more unusual rarities seen every year far from their natural range. Although the appearance of some vagrant birds can be explained by the accident of being swept off-course by contrary winds (see page 106), this is not the explanation for all such occurrences. The birds' appearance on a reciprocal course to that which would have taken them to their correct destination suggests that there has been a serious error of navigation.

Pallas's leaf warbler is an eastern Siberian species that makes regular appearances in western Europe. Its breeding range is in the great coniferous forests centred around Lake Baikal and its normal wintering area is in eastern China and neighbouring Indo-China. The explanation for individuals turning up as far west as the British Isles is that they have flown on a Great Circle route (see page 77–78) exactly opposite to the correct direction. Instead of flying about 4,000 kilometres (2,480 miles) they have kept going for 7,000 kilometres (4,350 miles). Most Siberian vagrants in Europe are young birds on their first migration. Some survive the winter and it is possible that a few get back home. Maybe the adult Siberian birds that are sometimes seen are birds returning to Europe for the second winter. The equivalent movement in North America is reflected in the records of blackpoll warblers and other North American woodland species that have turned up in Greenland when they should have migrated in a south or south-westerly direction. And it certainly must have been faulty navigation that brought a juvenile sandhill crane to the north-west corner of Greenland in 1985.

BELOW *The black-browed albatross is one of three species of albatross seen in the North Atlantic, far from their southern homes.*

PREDATION PRESSURE

The twice-yearly masses of migrating birds are a bonanza for predators, and there are many accounts of birds-of-prey taking advantage of the flocks passing by or roosting on their hunting grounds. The flocks also have to contend with birds-of-prey migrating with them. Observations on the passage of migrants at a site on the Baltic coast show that the migration of sparrowhawks coincides with that of flocks of bramblings and chaffinches. As much as 10 per cent of their numbers may be eaten *en route*.

The pickings are especially rich since the migrants are not only on unfamiliar ground and lack familiar escape routes to shrubs or other cover, but they are also fat and easier to catch because their flight is laboured. A steep take-off and climb to the cruising height is a good strategy for avoiding predators, but a fat warbler setting off on a long flight is so heavy that this must be difficult, so the birds with the best chances of flying the distance are at the greatest risk from predation. However, at the coastal reefs of the Banc d'Arguin off Mauritania, where the two million waders that mass there are preyed on by falcons of different species, observers have noticed that the smallest, and therefore weakest, birds are taken. Those that have so little flesh as to be near death are sometimes rejected by the falcons after they have killed them.

To combat the threat of predation, birds often take off together, so that each bird minimizes its chance of being the target of an attack, and they climb steeply to get clear of any birds-of-prey. I have seen gyrfalcons attacking formations of barnacle geese as they were flying low through valleys in Greenland on the start of their southward migration. On sighting a gyrfalcon, the geese closed ranks and flew in a tight flock. The gyrfalcon could attack only if a goose became separated from its companions. Once they had climbed a few hundred metres they were well above the patrolling height of the falcons.

Ornithologists watching radar sets to study migration across the North Sea noted that echoes showing the movements of birds disappeared from the screens during the early hours, then suddenly reappeared before dawn. The explanation is that the migrants dropped below the level of the radar beam, then climbed back into it. Scandinavian birds, mainly thrushes, approaching the British Isles lose altitude towards the end of the night, in an attempt to land before daybreak. They need to find safe roosts and avoid predators but they must also avoid overshooting into the Atlantic. If, however, they find themselves still over the sea when it begins to get light, they climb steeply to avoid patrolling predators, and perhaps also to search for a landfall.

As well as facing many other hazards on their journeys, small birds are harried by birds-of-prey that migrate with them, such as this merlin. For the predator, its accompanying source of readily available food saves the need to fatten up before the journey.

The interpretation of this behaviour came from observing migrants at oil-rigs in the North Sea. Low-flying birds, which may be suffering from exhaustion, are harried by herring gulls and great black-backed gulls which live around the rigs, and by predatory birds that are part of the migrating stream of birds: short-eared owls and long-eared owls by night and kestrels and merlins by day.

Attacks by gulls are often witnessed from boats. Their victims are nocturnal migrants which are still over the sea after daybreak, often because there is a head wind which has slowed them down; such a wind also forces them to fly low and makes them vulnerable to attack. The onslaught is started by one gull and others are quickly attracted to join in, swooping at the migrant, which takes evasive manoeuvres until caught in mid-air or forced into the sea. Some observers have noted that the victim is often abandoned uneaten. Perhaps the gulls are behaving like the proverbial fox in the hen run, in that their hunting is being stimulated by a glut of prey that cannot escape, although they are not hungry.

Two birds-of-prey specialize in catching migrating birds by delaying their nesting until the autumn passage provides them with an abundance of prey on which to feed their nestlings. Eleonora's falcon nests from the Canary Islands to Crete, and the sooty falcon nests in the eastern Sahara and Arabian Peninsula, both in the desert and on islands. Both species lay their eggs at the end of July and have growing young to feed between the end of August and the beginning of October. After breeding they migrate mainly to Madagascar and nearby islands, where they live in flocks.

Eleonora's falcon nests in small colonies on cliffs and catches nocturnal migrants that are still flying over the sea after dawn. It is essentially an insect-eater and pursues birds only during the breeding season. It does not have the usual adaptations of a bird-hunter (long legs, large feet and talons, and high wing-loading) and relies on the vulnerability of its prey. The passing migrants cannot land and are exhausted by successive attacks, but many still get past the falcons and fly on to safety.

The hunting success of each pair of falcons, and so also their breeding success, depends on their location. If there is land nearby to the north, where the migrants have been able to alight by daybreak, there will be few birds still crossing the sea in the morning and hunting is poor. On the island of Mogador, off the coast of Morocco, however, the falcons catch birds which have taken off from Portugal, 600 kilometres (370 miles) to the north, and many more are airborne through the day on the long sea crossing. Eleonora's falcon has a total population of about 5,000 pairs. They kill some two million small migratory birds, mainly warblers and shrikes, each autumn, but this is out of the enormous total of about 5,000 million landbirds migrating to Africa, and has no significant impact on the species involved.

HUMAN THREATS

Flocks of migrating birds have always been regarded by man as a bonanza of fresh food, whether of wheatears snared by English shepherds (see page 42), Eskimo curlews shot by American farmers (see page 44) or quails caught in Italy. (The Bishop of Capri received much of his church's taxes from the annual harvest of quails.) From serving as a renewable resource to feed the human population, migrants are now trapped and shot for sport or the supply of luxury foods. The legality of this killing is based on the earlier necessity but it is now often carried on at levels beyond the sustainable yield, and every passing bird is considered a fair target.

In Europe awareness of the plight of migrant birds has been focused on the slaughter of birds as they fly through countries bordering the Mediterranean Sea. No bird is allowed to pass: honey buzzards, falcons, herons, doves and all kinds of songbirds are netted, trapped, shot or stuck fast to perches with bird-lime. It is calculated that hundreds of millions are killed each year, increasingly for sport rather than to earn a living. Bird-lovers in the north of Europe are particularly incensed because these are the birds which should be filling their own woods, fields and gardens in the summer.

In 1981, the Directive on Conservation of Wild Birds came into force in the European Common Market. Among other

ABOVE *Huge numbers of European migrants, like this song thrush, are trapped or shot as they fly through Mediterranean countries.*

BELOW *Oil pollution is a threat to migrant birds at their staging posts. It destroys food supplies and ruins the plumage of birds that come into contact with it.*

Development, in this case a marina, can permanently ruin a staging post and perhaps remove an essential link in a migration route.

measures, it requires member states to ensure that migratory species are protected in spring. However, the hunting lobby in many countries is more powerful than the conservation lobby and the Directive is frequently ignored. In 1984, the Greek government actually repealed a law banning shooting.

Even where migrants are safe from hunting and shooting, or these activities are at least controlled, their numbers are threatened by development. The ribbon of building along the coast of the Gulf of Mexico threatens migrants that pitch onto the shore to recuperate after a hard crossing from the south.

Modern warfare can also have serious consequences for migrant birds. The huge quantities of oil released after the deliberate destruction of Kuwait's oilfields during the Gulf War may have long-lasting effects on migrants passing through the Persian Gulf. Many waders and others died from landing on oil pools and slicks, and many more have no doubt been contaminated sufficiently to die after they have flown on.

Among the most vital staging points are wetlands and estuaries which provide rich feeding and safe shelter for many migrants. These habitats are particularly favoured by developers. The loss of marshes and pools to agriculture is a problem for migrating wildfowl, as is the conversion of the shores, saltmarshes and mudflats of estuaries that are major staging posts for waders. Pollution is an additional burden. Without these vital links in the migration network, entire populations of birds will be put at risk, and the preservation of staging posts has become an important part of the bird conservation movement. Britain has more estuaries than any other European nation, although one-third have been reclaimed since Roman times, and the process continues even in estuaries designated under the Ramsar Convention on Wetlands of International Importance.

When migrants arrive at their winter homes they face more threats from persecution and habitat destruction. While most people are familiar with the wholesale ruination of tropical habitats, it is not so generally appreciated that this also affects the wildlife of northern countries. The severe droughts in the Sahel region immediately south of the Sahara Desert, for example, have adversely affected migrants. The bushes that survive the drought and which would offer food and shelter to migrating birds are lopped to provide greenery for livestock. The result is fewer birds singing in Europe next spring.

The situation is even worse in America, where so many migrants spend the winter in the tropics. The dwindling numbers of wood warblers, flycatchers, vireos and other songbirds in the forests of the United States and Canada are linked with the destruction of forests in Central America and northern South America. Many species require mature stands of trees as a winter habitat but these are being logged and converted to grassland. No matter how hard North Americans try to make their own forests safe for nesting migrant songbirds, their efforts are bound to be doomed to failure if the families the birds raise fly off into oblivion.

CHAPTER 6

WINTER HOMES

The traditional view of migratory birds in northern countries is that they fly away to warm countries to escape the cold, barren winter season, and then reappear in spring when the weather improves. Little thought is given to what they do in their winter homes. Most scientific research has concentrated on the departure of birds and on their journeys. Only recently has attention turned to the details of how they live in their 'second homes', perhaps half a world away in very different conditions of climate and habitat, and where they may spend more than half of each year.

White storks roost on a dead acacia in East Africa, far from the roof-tops of the European town where they nested.

WINTER LIFE

Life in the winter home is just as challenging for a bird as it is in its summer home. The marvellous ability of a young bird to navigate to a precise spot on the Earth's surface is only half the story of migration. It arrives in a strange place, which may be very different from the home it grew up in, and it is faced with the same problems as it had when it left the nest. It must find plentiful sources of food and a safe place to roost, and it must learn to avoid a new set of predators. It may also be hampered because it is trying to take up residence in a place that is already occupied by the native bird population and the adults of its own species.

From the viewpoint of Europeans and North Americans, the majority of migrant birds 'live' in the north temperate zone where they nest, and their flight south is to temporary accommodation where they subsist while their northern home is uninhabitable. Yet it can be argued that these birds should be regarded as tropical species that fly north to take advantage of a seasonal abundance of food for breeding. There are, for instance, eight species of swallows in southern Africa, so should not the familiar 'European' swallow be regarded as an African species that visits northern lands for the summer? The argument for migrants being natives of their southern home is even stronger for the hummingbirds, which are such well-known inhabitants of the New World tropics. Of the total of about 320 species, only five migrate into temperate latitudes, with the rufous hummingbird reaching as far as Alaska.

Some perspective can be given to the question of where a bird's true home lies by looking at the migrants' calendar. American warblers spend only three months of the summer nesting in Ontario, Canada, but six or seven months wintering in the tropics. American migratory hawks spend four months in the north, four in the south and four on migration. On the other hand, when a bird stops migrating and remains permanently on its breeding ground, it is reasonable to conclude that this is its true home. This appears to have happened with 'European' sand martins and white storks which have established breeding populations in their South African 'winter' homes, and 'North American' barn swallows which are now breeding in Argentina.

It can also be argued that the division into summer and winter homes may be an unnatural distinction and that the birds should be considered as having one large range over which they roam throughout the year. The idea of a 'summer home' and a 'winter home' as fixed points can be misleading because of the complexity of many migrations, in which birds progress through a series of stopover sites and continue a nomadic existence on the wintering grounds. They may be stationary only when tied to their nests, and this need not be in the same site or even locality each year. Crossbills, for

The yellow warbler nests in many North American wooded habitats, then migrates to a very different winter home in tropical America.

instance, nest where there are good crops of conifer seeds, and individuals may move many kilometres between nesting seasons to find a good crop.

The continuous nature of migration is illustrated by American thrushes which spread slowly through Central America, remaining for several weeks at some places, and gradually move south into South America. By the time they have reached the southern part of their range, it is time for them to turn round and work their way north again. However, the nomadic behaviour of birds in their winter homes is easier to appreciate with familiar winter visitors to temperate latitudes.

It is a common observation that winter visitors may hardly be seen for weeks and then turn up, overnight, in large numbers. In Britain, the two such birds most noticed are the fieldfare and redwing, thrushes which breed in Scandinavia, with some also from Iceland in the case of the redwing. (Scandinavian blackbirds, song thrushes and mistle thrushes also come to the British Isles in winter but they are indistinguishable from the natives.) The fieldfare's name is Anglo-Saxon in origin and means 'traveller over fields', showing that our ancestors were struck by its appearance in winter: Chau-

When the redwing first arrives in western Europe from the north in autumn, it feeds on newly ripened berries. Flocks wander around the country in search of good crops, which they quickly strip before moving on.

cer referred to 'frosty fieldfares' in his poems. Fieldfares and redwings are nocturnal migrants and their calls – the harsh chuckling of fieldfares and the thin whispering of redwings – can be heard overhead at night. During the day, conspicuous flocks can be seen settling in fields and hedgerows.

When they arrive in autumn, fieldfares and redwings are searching for crops of wild fruit – hips and haws, holly and rowan berries – and they come into gardens for scarlet berries of cotoneaster and pyracantha, and for windfall apples. They have to move when stocks are used up and eventually they are forced into the fields in search of worms and other small animals. When frost locks up the soil, the flocks join the general exodus of birds, generally westwards into the milder south-west of Britain and Ireland, or across the sea into France and Spain.

The randomness of these movements is demonstrated by the recovery of ringed birds in widely different places from year to year. Redwings ringed in winter in Britain have been recovered in subsequent winters as far away as Italy and Iran. The reason is probably the uncertainty of fruit crops. Sticking to a rigid itinerary is possible only if food supplies are reliable. Unless the area has been devastated, a swallow can return to its winter home on the savannah, or a wader to an estuary, with confidence that it will find plenty to eat. Fruit-eaters, by contrast, must be prepared to travel in search of food because fruit crops are liable to fail simultaneously over a wide area.

PACKING THEM IN

It has been estimated that about 5,000 million landbirds leave Europe and Asia each year and travel into sub-Saharan Africa. This figure must be combined with a large but unknown number of waterbirds such as ducks, waders, gulls and others. Given that Africa is already well endowed with its own resident birds, there would seem to be a difficulty in fitting all these newcomers into the available space. There are similar problems in other tropical regions of the world; another estimate is that half the migrant landbirds from North America fit into an area of Mexico and the West Indies equivalent to one-eighth the area of their breeding grounds. Ornithologists have, in recent years, been paying more attention to the winter ecology of temperate breeding birds, previously a missing dimension in the study of their life-cycles.

The relevance of the ecology in distant lands to a country's breeding birds was brought into sharp focus when the British population of the whitethroat 'crashed' between two breeding seasons because of excessive drought in its African winter home, as shown on the bar graph on this page. On the other side of the Atlantic, there is great concern in North America,

The fate of the whitethroat serves as a lesson that the conservation of birds is an international problem.

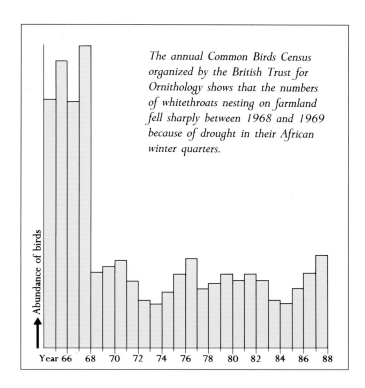

The annual Common Birds Census organized by the British Trust for Ornithology shows that the numbers of whitethroats nesting on farmland fell sharply between 1968 and 1969 because of drought in their African winter quarters.

Abundance of birds

Year 66 68 70 72 74 76 78 80 82 84 86 88

where the disappearance of common birds is linked with the destruction of their winter habitats in tropical forests (see page 115). Temperate wetland habitats which are vital winter homes for Arctic birds are also under threat (see page 126).

There is no easy way of describing how migrants fit into their winter homes because the situation varies according to species and, within a single species, to different places. For instance, bay-breasted warblers feed on insects and defend territories when they arrive in Costa Rica, but gather into flocks and feed on nectar and fruit when these become plentiful later in the winter. Sometimes, territorial individuals live a few hundred metres from others living in flocks. On the other hand, chestnut-sided warblers living in the same region always hold territories, no matter how abundant their food.

The differences in the behaviour of individuals of one species has been studied in detail at Teesmouth, a river estuary in north-east England, where colour-ringing shows that grey plovers (known as black-bellied plovers in North America) have several strategies for wintering. When juveniles arrive from the breeding grounds, they may spend the rest of the winter on the estuary, or they may move on to other sites after a few weeks. Those that stay try to gain a territory but some are unsuccessful because of competition from established adults, and so they lead an unsettled life wandering around the estuary. In future years the plovers maintain the same strategies that they adopted in their first winter, and individuals that established territories return to exactly the same patch.

The draining of birds in autumn from the huge landmasses of North America and Eurasia towards the tropics and beyond leads not only to a concentration of the migrants but also to a tight fit with the residents that are already there. Avoiding competition with the residents is a major problem for migrants. This may be resolved in several ways. The migrants can squeeze in, being able to occupy less 'room' because their nutritional requirements are lower than in the summer. They no longer have the demands of rearing families and the warm weather reduces the energy they need to spend to keep warm.

Local populations of residents may be limited in numbers by a reduced food supply at a time when the migrants are absent, so there will be too few residents to take full advantage when food does become abundant. Thus when the migrants arrive, the two populations share the abundance without too much competition, as happens among 44 species of hummingbirds studied in Mexico. In summer the crop of flowers limits the numbers of resident hummingbirds, and the main flowering season starts when the 10 species of migrants arrive, ensuring there is plenty for all.

Competition is further reduced if migrants find ecological niches – places where they can live – that are empty of competitors. The numbers of birds-of-prey that enter Africa

The common crossbill's winter distribution is determined by the abundance of the crop of conifer seeds, but the birds are sometimes forced to feed on other seeds, such as thistle seeds, as seen here.

ABOVE *The behaviour of a grey plover in winter depends on whether it manages to acquire a territory on the shore. If it fails, it will lead a nomadic life until it is time to fly north.*

from Eurasia exceed the residents, but nearly all occupy open savannah woodlands and grasslands, where they tend to take up different ecological niches from the residents. The four species of harriers that come from Eurasia have virtually no competition for their lifestyle on the African grasslands, because the two native species are very restricted in their range. A similar situation has been found with northern waterthrushes and American redstarts wintering in Venezuela. When these species migrate back to North America, their niches are left vacant, so there is no 'crowding out' of residents by migrants when they return. Ruby-throated hummingbirds wintering in Costa Rica either use nectar resources

that are not greatly exploited by resident species or they join in feeding on local but very abundant supplies that the resident species cannot fully utilize.

One factor that helps the migrants is that the residents may be tied to one place by nesting duties when the migrants are present. The latter can then live as opportunists, moving to take advantage of temporary sources of food, such as animals fleeing into the open from bush fires.

HOW FAR TO GO

The inbuilt navigation programme described in Chapter 4 takes a bird to a broad area where, over generations, its predecessors have found favourable habitats for wintering. Unless a young bird has accompanied its family or travelled with a flock of its fellows, it is faced with the need to find a place to settle. This will be in a habitat that may be very different from the one it left behind. Having flown south to approximately the correct wintering area under the control of the rigid, instinctive programme demonstrated in the *Zugunruhe* experiments on captive birds (see pages 68–69), the migrant must switch to flexible behaviour so that it can search for somewhere to settle and adapt to local conditions.

BELOW *The first snows drive the green-winged teal out of the prairies and south to Mexico, but in mild winters some remain as far north as southern Alaska and Newfoundland.*

Some evidence for this flexibility is seen in species that do not always return to the previous year's wintering site. Some ducks and waders spend winter as far north as conditions will allow and move to the end of their range only if pushed by cold weather. Common cranes are reluctant to move from stopover sites while food remains and they will overwinter in France unless hard weather forces them over the Pyrenees.

Staying as far north as possible in winter saves mileage on outward and return journeys, while the choice of moving further has a survival advantage. Yellow-rumped warblers have been found to maintain a physiological readiness to continue migrating through much of the winter. Some winter in Arizona deserts where they find insects along watercourses. If cold weather spreads down from the north, the insects disappear and the warblers reactivate their migratory behaviour and head further south.

WINTER WADERS

Estuaries are vital for the survival of many species of waders and other birds, either as staging points on migration or as winter homes. The daily ebbing of the tide uncovers broad tracts of mud and sand. Here the birds probe and pick, each species in a different way, sharing the wealth of invertebrate animals that thrive on the rich organic matter brought down by rivers and deposited at the edge of the sea. Only high winds, an impenetrable layer of ice or disturbance by human activities will prevent them from feeding. Estuaries are wild places and the flocks of waders that spread out to feed and gather in immense flocks to fly inland as the tide returns are among the most memorable sights in nature. Both estuaries and their birds are under threat, however (see page 115). Conservation of estuaries is therefore a priority for the survival of wader species.

It is estimated that of the 3¼ million waders wintering on the Atlantic coast of Europe, about 80 per cent gather in no more than 38 sites, with the majority in the Wadden Sea of Germany and the Netherlands, and large numbers in Morecambe Bay and the Wash in England. Three decades of studies at these major sites have demonstrated that there are complex movements of birds through the winter.

The most abundant wader wintering in western Europe is the dunlin. Well over a million individuals arrive each year from northern Russia and Scandinavia, but many more from Iceland, Greenland and the temperate parts of Europe pass through on their way to Africa. A British study of the recoveries of ringed birds and sightings of dye-marked birds has given a picture of the winter life of this little wader.

When dunlins first arrive in British estuaries from their northern tundra and moorland breeding grounds, from July onwards, they become fairly sedentary. The adults spend this time moulting and the juveniles probably finish their growth and practise feeding techniques in their new habitat. At the end of the moult, in October and November, some birds stay in the same place but there is a considerable movement, mainly in a westerly or southerly direction. The Wadden Sea loses dunlins which cross to the British Isles or fly down to western France, Spain and Portugal. The Wash also loses dunlins but other sites fill up at this time. The following midwinter period has little movement, but from mid-February onwards the dunlins are moving again and numbers in the Wadden Sea and Wash rise. These are the areas where the dunlins will moult into their brighter breeding plumage and put on weight for migration.

A flock of dunlins gathers in a crowded roost at high water. When the tide falls, the birds will spread out over the shore to feed among other species of waders, sharing the resources.

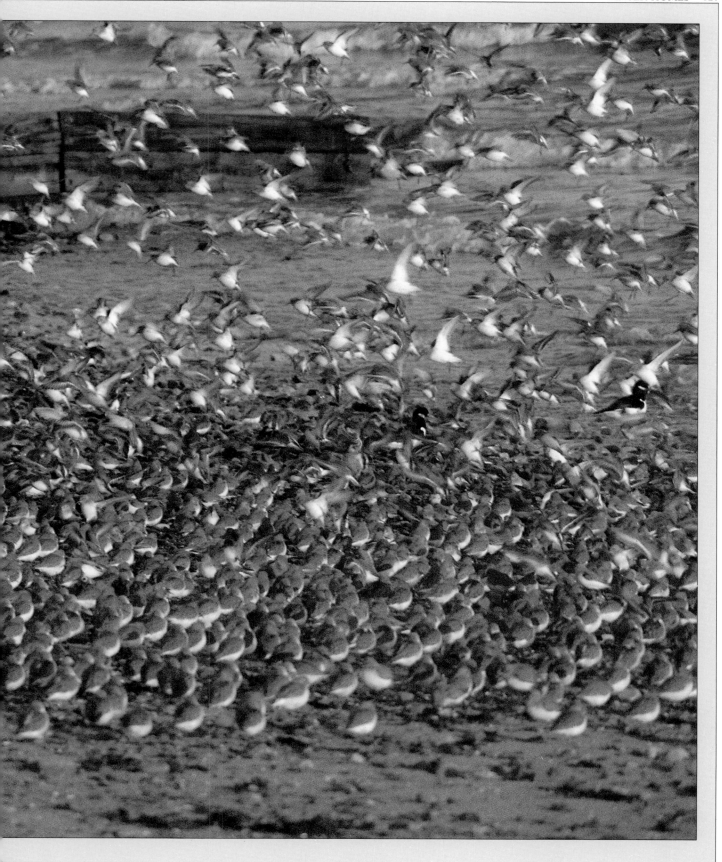

MOVEMENTS IN THE TROPICS

Nearly 70 years ago, a British resident in Egypt noticed that a white wagtail with only one leg spent two successive winters in her tiny patch of garden near Alexandria. Here was proof that migrant birds were as faithful to their winter homes as they are to summer nesting sites. This was soon confirmed by pioneer ringing studies in Africa. One yellow wagtail was caught in the same place in Nigeria for seven consecutive years. It was a fairly exceptional bird to have lived so long, and many more yellow wagtails have been caught in the same place over two or three seasons. There are similar records for swallows and other species.

Some of these wintering birds set up territories in the tropics, which they defend by singing, so giving bird-watchers from the north the experience of an incongruous mixture of familiar bird-song in an exotic setting. The marsh warbler sings for a longer period in Africa than it does in Europe. In tropical America, some of the wood warblers, such as chestnut-sided, magnolia and yellow warblers, defend one territory throughout their stay in winter quarters, but other birds hold a series of territories as they move around the winter range. Garden warblers arrive in Uganda during November, hold a territory for six weeks and then move on again, presumably heading southwards. They return in February on their northward migration for another stay and take up exactly the same territories. Their fidelity is not surprising because returning to a familiar place enables them to save time and energy.

Holding a territory is sensible only if there is a fixed resource to defend and many visitors to Africa remain nomadic through the winter as they follow shifting food supplies. Birds-of-prey entering Africa keep on the move to take advantage of the appearance of great swarms of termites and locusts or the huge flocks of the little weaverbird called the red-billed quelea, whose own movements are determined by the shifting rains. Their search for prey leads the hunters to follow the rain belts and, as the British ornithologist

BELOW *When it reaches Africa in a few months' time, this garden warbler will sing again as it defends a winter territory in a very different landscape to the European oak wood in which it breeds.*

Ian Newton put it: 'In a sense they ride on the crest of a wave, continually shifting from one temporary abundance of food to another'.

— AFRICA —

On a continental scale, migrants to Africa shun the block of rain forest centred on the Congo Basin. The orioles are the only songbird family to winter exclusively in rain forests but swallows, pied flycatchers, hobbies, nightingales and a few other birds can be found in the forest zone, at least while passing through. Most migrants favour the open woodlands and grasslands of the surrounding savannah regions. Probably the majority are to be seen in the harsh, dry Sahel bordering the Sahara Desert – the opposite of what might be expected.

The geographical advantage of the Sahel is that it is not so far to fly to from breeding grounds in temperate countries. However, it has the appearance of being very inhospitable. (Arrivals from the Arctic, in particular, are faced with adapting to an enormous increase in air temperature and solar radiation.) The sensible strategy would seem to be for the migrants to continue further south to reach moister, lusher habitats. This has become known as Moreau's Paradox, after the pioneering British ornithologist Reg Moreau. In his classic book on the migration of birds into Africa, Moreau points out that the worse a landscape seems to be for birds, the greater the number of species of migrants that inhabit it. More species of insectivorous birds overwinter in the dry savannahs than elsewhere, and the greatest numbers are found in the notoriously dry region of the Sudan and Ethiopia. This region is a parched waste from the end of its scant rainy season until the following year's rains, if indeed they do arrive. Yet the migrants not only survive but also find enough food to build up reserves for the spring flight.

The resolution of this paradox is based on the patterns of rainfall in sub-Saharan Africa and their effect on the continent's ecosystems. The rainy seasons are caused by moist, oceanic air masses blowing westwards into the continent, where they warm and lift into the atmosphere. The air cools as it rises, then its moisture condenses into clouds and falls as rain. The belt of wet weather caused by this air movement swings like a pendulum to and fro across the equator, bringing rain to the north in the northern summer and to the south in the northern winter. The rain is heaviest and most prolonged around the equator, where there are two rainy seasons, and becomes sparse and abbreviated towards the northern and southern borders of the tropics.

The rains reach the Sahel, which merges with the southern boundary of the Sahara Desert, in a short season from June to September. The bulk of the migrants arrive in September and October when there is a flush of plant life and the insects that feed on it. Most of the migrants are insect-eaters (turtle doves and quail are among the few exceptions), and for two or three months there is plenty to eat. However, even in the dry season there is lush growth in wet swamplands and river valleys, and some of the Sahelian trees and shrubs continue to sprout leaves through the drought. They also bear fruit during this season and the insectivorous birds switch to eating berries, which provide the energy needed to lay down fat for the northward migration in spring.

Others migrants use the relatively good conditions that prevail in the Sahel after their arrival to fatten up and continue their migration southward, around the eastern edge of the Congo forest towards the southern savannah zone. The scale of this movement was realized when a British ornithologist, Alec Forbes-Watson, visited the tourist lodge in the Ngulia Hills at the centre of the Tsavo National Park. One of the attractions of the lodge is the use of floodlights to illuminate the animals coming at night to a waterhole in the grounds. One misty, wet night in 1969, Forbes-Watson noticed large numbers of birds gathering in trees and bushes around the lodge. The floodlights were attracting passing migrants in the same way as a lighthouse.

In the next 10 years over 30,000 birds were ringed at Ngulia, mainly from mid-November to late December, and they included over 40 species from Europe and Asia. Since then, the broad pattern of bird movements within Africa has been revealed to a considerable extent, although the details are very sketchy compared with what is known about the European end of the migration route.

The turtle dove migrates from Europe and Asia to winter in the Sahel region of Africa and further south. Huge numbers gather in roosts on the floodplains and descend to feed on spilled grain.

After the migrants' arrival south of the Sahara, they stop for one to three months and feed to regain some of their lost weight. As the habitat dries out, they move on, following the southward drift of the rainy season, through the months of November and December. They then move on southwards to Uganda and South Africa. The sedge warbler does not arrive at Lake Uganda until February, long after it left its summer home. This is only a month before it starts its return and emphasizes that migration is a continuous movement rather than a straightforward out-and-back commuting between a species' summer and winter homes.

— AMERICA —

The pattern of migration in the Americas is rather different from that seen in the Europe and northern Asia/Africa system. In contrast to the Old World migrants' shunning of the Congo rain forests, American migrants are more common in both rain and dry forests of tropical America. Some inhabit solid, untouched forests while others are found around the edges or in secondary growth.

Another difference is that the Old World birds are faced with the crossing of the inhospitable Sahara Desert whereas American birds can settle in the rich habitats of Central America. The majority of North American migrants go no

ABOVE *Wintering white storks stride among grazing zebras in the Ngorongoro Crater, Tanzania. They gather in large numbers to feed on swarms of locusts, caterpillars and other insect pests.*

BELOW *The winter distribution of the European wood warbler is confined to a region of Africa north of the Equator. There, it undergoes a complete moult for the return flight and breeding season.*

further than Central America and only a minority reach the northern countries of South America. In contrast to the situation in Africa, few migrants cross the equator and reach southern South America. The exceptions include barn and cliff swallows, bobolink, Swainson's hawk and several waders.

In Mexico, migrants may make up half the winter bird

BELOW *The bobolink makes the longest migration of any member of the American blackbird family. It travels from its breeding grounds in the grasslands of North America to marshes and ricefields in southern Brazil and Argentina.*

population but they represent only a small fraction of the incredibly rich avifauna in the Amazonian basin. This is seen as a clear trend in the wood warblers, the largest group of migrants. Migrant species outnumber their resident ecological counterparts in southern Mexico, Guatemala and Panama, but the situation is reversed in Surinam and Bolivia. The migrant populations of songbirds in Central America tend to fill vacant ecological niches and defend territories, while those in South America tend to be subordinate to the residents and survive the winter through the opportunistic use of temporary surpluses of food, such as army-ant swarms.

AMERICAN WOOD WARBLERS

Roger Tory Peterson, the renowned American ornithologist, described the wood warblers as the 'butterflies of the bird world' because of their bright colours. Sometimes called American warblers to distinguish them from the unrelated Old World warblers, which confusingly include a species known as the wood warbler, they are a group of about 130 species of songbirds that are favourites with birdwatchers. Some are common and easy to recognize; others are rare and extremely taxing to identify.

Wood warblers are fascinating for their variety of lifestyles and the waves of colourful little birds migrating through woods and towns provide an exciting spectacle in spring and autumn. Of the 51 species that nest in North America, all but the rare tropical parula of the Rio Grande valley, Texas, are migrants. Although a few remain in the southern United States all year, chiefly in the south, 47 species join their relatives in Mexico, Central America and northern South America in the winter, and form the largest group of bird migrants to enter the American tropics.

When the warblers reach their winter homes, where they will stay at least twice as long as in their summer homes, they enter a world very different from the conifer forests, hardwood stands, willow-scrub bogs and other habitats in which they were reared. Even those yellow-rumped warblers that remain in North America, rather than migrating further south, move from their coniferous and mixed-forest nesting habitat to a much wider range of habitats for the winter. Moreover, migrant wood warblers are funnelled into a smaller land area in the tropics than they occupied during the summer in temperate North America. The congestion is increased because the region is already filled with many tropical relatives, as well as similar but unrelated types of insect-eating birds.

North American ornithologists familiar with warblers on their breeding grounds have travelled to tropical America to see how 'their' birds fare in winter. They have found that competition with tropical birds is reduced because most of the immigrant warblers settle in Central America and northern South America, whereas the residents are concentrated further south. Where the two groups do meet each other, migrants and residents tend to differ in feeding places and habits. For instance, migrants tend to feed among the foliage while residents feed in the undergrowth or hawk for flying insects. Furthermore, migrant warblers also avoid competition with each other, to the extent that male hooded warblers settle mainly in primary, virgin forest while females prefer secondary, regenerating forest.

The general conclusions are that the winter season may be the critical time in the wood warbler life-cycle but that the migrants should not be regarded as invaders that have to force themselves among the residents to glean a living. They are essentially tropical birds, well adapted to life in their winter home, and they fly north in spring to take advantage of the relatively empty habitats of North America. While they are away, their places in the habitat are not usurped by the residents but are left empty until their return. So it seems that there is no overcrowding in tropical America due to the arrival of North American birds that have come to escape the northern winter. Rather, there is a temporary thinning as birds leave to exploit the northern summer.

LEFT *The yellow-rumped warbler is one of the most familiar American warblers because, unusually for a wood warbler, it winters in North America, where it is a frequent and confiding visitor to bird-feeders.*

LEFT *The breeding range of the American wood warbler family compared with the smaller area of the family's mainly tropical wintering range.*
BELOW *The life cycles of three representative species of wood warblers, showing the time spent on the breeding and wintering grounds and on migrating between the two. They stay longest in their winter homes.*

■ Breeding areas

■ Wintering areas

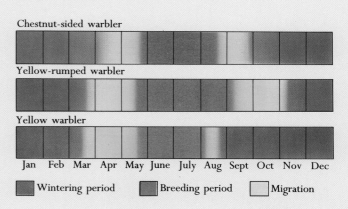

Chestnut-sided warbler

Yellow-rumped warbler

Yellow warbler

| Jan | Feb | Mar | Apr | May | June | July | Aug | Sept | Oct | Nov | Dec |

■ Wintering period   ■ Breeding period   ▢ Migration

CHAPTER 7

THE RETURN

The return to the breeding grounds in spring is like the southward journey in autumn in reverse, but there may be important differences in route as the birds take advantage of prevailing winds and staging posts. The imperative to breed drives the birds to fly north as soon as possible so they can make an early start to nesting as soon as conditions on the breeding grounds are suitable.

Flocks of snow geese stream in formation over California on the start of the long journey back to their Arctic breeding grounds.

THE RACE TO GET BACK

Niko Tinbergen, the Dutch zoologist who, in 1973, won the Nobel Prize for his research on animal behaviour, once spent a year in Greenland. He recounts how, on 22 March, in the middle of a blizzard, a young Greenlander came tumbling into his house, calling excitedly 'The snow buntings are coming'. Outside, they found three small birds sheltering behind a stone. The return of the snow buntings signalled that winter was over; for the Greenlanders their appearance was an even more significant event than the arrival of the swallows that heralds spring in temperate regions.

Tinbergen could not at first understand how the snow buntings would survive in the snow-covered, frozen landscape but, later, he was to observe that, for a month, they subsisted on seeds of grasses protruding through the snow, until the thaw exposed enough ground for the males to stake out territories and start courting. It is important for the buntings to return as early in the year as possible so they can start breeding and rear their families before the end of summer.

This is more difficult for birds to achieve in the Arctic (or Antarctic) than at lower latitudes with longer summers, but at all latitudes there is a race to start nesting.

Many studies have shown that early nesting, even by resident birds, produces the most young, so it is not surprising to find birds returning 'hard on winter's traces'. In northern Europe, comparison of records shows that the first arrival date of several species has become earlier as spring air temperatures have risen and plants have started to grow earlier over the course of the twentieth century.

RIGHT *Common cranes start courtship while on migration and arrive in pairs at their breeding grounds. These birds at a staging post in southern Sweden are waiting for the weather to improve.*

BELOW *The arrival of the snow bunting is a sign that the Arctic winter is over. The cheerful song of the boldly plumaged male can be heard over landscapes still covered in snow.*

ON THEIR MARKS

As with the autumn migration, preparation for the spring migration starts several weeks before departure. The hosts of songbirds that spend the winter in the Sahel (see pages 129–130) have to fatten up during what appears to be the most unfavourable season of the year. To some extent they gather around the more verdant shores of Lake Chad or other wetlands, where there are swarms of midges, but many of these insectivorous birds turn to a fruit diet.

Waders that winter on the coastal reefs of the Banc d'Arguin, off Mauritania, start to put on weight in anticipation of their flight north up to six weeks before departure. Throughout the winter the time spent feeding has been related to their weight: if they were lean, they fed more until they fattened, then the intake was reduced. When preparing for migration, however, this relationship is lost; the waders increase feeding time and their weight increases beyond the winter norm. They start to feed at night and continue through neap high tides when they would normally rest, and their rate of feeding also increases. Whimbrels, for instance, can feed more rapidly because, at this time, the crabs that form their main prey become more active and so are more easily caught.

The date of departure may be controlled either by the birds' internal biological clock or by their external environment. Birds wintering in temperate regions can use a predictable lengthening of the day and improvement in the weather as cues for starting the migration process. Those wintering in the tropics may rely more on their biological clock as a timer, because there is little change in daylength to provide an external cue about the passage of time, and also because rainfall and temperature changes are unpredictable from year to year.

Analysis of the dates of passage of a wide range of songbirds at the Manomet Bird Observatory on Cape Cod Bay, Massachusetts, shows that birds wintering in southern North America reach the observatory, on average, 17 days before those wintering in Central and South America. It seems that the former, such as the yellow-rumped warbler, can cope with harsher conditions, so they press northwards hard on the heels of the retreating winter, while the latter are reluctant to leave their pleasant tropical environment. When they do move, there is a concentrated passage of birds past the observatory as they race to the breeding grounds. The blackpoll warbler, which winters at the far south of the wood warbler family's range in South America, also nests at the far north of the range in North America. It migrates in spring after other species because its summer home is late to clear of snow, but on its final flight through Canada and into Alaska it covers over 300 kilometres (186 miles) per day, compared with most wood warblers' usual 50 kilometres (31 miles).

Before starting its long trek north, the whimbrel feeds to put on fat to fuel its flight. It is helped if it can find rich sources of food so that it can forage rapidly and efficiently.

GEESE TO THE ARCTIC

The geese that I watched in spring gathering at Cape Wrath (see page 53) were in a hurry to fly north. It is to their advantage to set off from Europe and head for their breeding grounds in Iceland or Greenland as soon as they are ready to go. This is not possible if the preceding month has been cold. The geese rely on a fresh growth of grass to fatten up for the journey, so a late spring holds them up. As the season progresses, the need to depart becomes more urgent because the geese must get to their Arctic nesting grounds to make the most of the short summer. They become increasingly likely to set out in less favourable weather and battle against headwinds to make up lost time.

Brent geese (known as brants in North America) return to the Arctic before the spring thaw has uncovered their food. If they do not start nesting as soon as possible, neither they nor their goslings will be in a fit state to fly south before the Arctic winter returns. When they first arrive, the geese may find some exposed vegetation and the female feeds furiously, while the male stands guard, but most of the

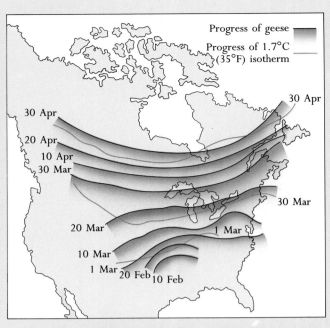

BELOW *Snow geese fly over the frozen sea near the end of their spring migration. They still have a reserve of fat which will sustain them until the thaw uncovers their food plants.*

ABOVE *The northward spread of Canada geese through North America corresponds to the improving spring weather, as shown by the movement of the 1.7°C (35°F) isotherm.*

energy needed for the formation of the eggs has to come from the fat store the female has brought with her from the winter quarters. Dutch ornithologists have found that they can predict the breeding success of female brent geese nesting on Siberian islands by watching their rate of feeding on the Dutch marshes in spring. They then check their predictions against the number of goslings each female has in tow when the geese return to the Netherlands in autumn.

The favourite spring food of brent geese is a plant called sea plantain and they fight for it. The females that get the most to eat are those with dominant mates who defend the best patches of vegetation. They put on large amounts of fat and the heaviest will lay the most eggs. Also, the fatter the

Barnacle geese flight in to feed at Svalbard. The thaw has uncovered the sparse Arctic vegetation so they can rebuild the reserves depleted during migration and the early part of breeding.

female, the less time she leaves the eggs uncovered, as she does not need to do so much foraging. The result is that the female geese that are fattest when they leave the Netherlands come back with the most goslings. A store of fat is also important for the male. The reserve of food allows him to spend more time guarding first the eggs and then the goslings. (There is evidence from other species, such as the swallow, that the link between pre-migration weight and breeding success may be a general rule.)

EARLY RETURNS

To get back and start nesting early requires a different strategy for migration compared with the more leisurely approach in autumn. The wood warbler of Europe (a species not to be confused with the American wood warbler family) takes 60 days to fly from Germany to Africa in autumn but only 30 days to return in spring. While an early return to the nesting place gives birds a head start in breeding, they run the risk of starvation if there is a return of winter weather. Arrival time is probably a balance between the advantage of an early start and the occasional death toll in exceptional weather, or at least poor breeding as a result of it.

Some waders – knots, turnstones and sanderlings – that nest in the high Arctic miss some staging posts on their way north in spring. They gather at a few major sites (such as Delaware Bay in the United States or the Wadden Sea in Europe), where they put on considerable weight, and fly over other possible sites without stopping. One reason for this is that the food supply at the latter places may be poor at this time of year. The shores of the Bay of Fundy in Canada and the Baltic Sea in Europe, which are important staging posts in autumn, are visited by few waders in spring because winter ice will have killed many of their intertidal prey. Another reason may be that the waders are in too much of a hurry to stop.

The strategy for any migratory journey includes a compromise between saving energy and time. An economical journey is made in a series of short, leisurely stages so that a bird does not have to carry a heavy load of fuel. In the spring, the trend is to travel fast, saving time at the expense of economy. The bird can save time on its northward journey by cutting out some of its refuelling breaks. The drawback is the extra fat it has to carry; this makes flight more strenuous and increases the rate of fuel consumption. Although current knowledge of fuel loads and flight range is far from complete, it is likely that the Arctic waders described above are abandoning economy in the race back to their tundra breeding grounds.

There is a visible northward spread of some common birds as the weather improves. It is quite correct to say that a single swallow does not make a summer; it is the arrival of the main body that shows that the weather is inexorably warming, with enough insects in the air to sustain substantial numbers of swallows. Their movement from Africa is linked to climate. After their 'forced march' across the Sahara, swallows travel at a more leisurely pace through Europe. Their passage parallels the northward advance of the 8.9 degree Centigrade (48 degree Fahrenheit) isotherm and they move at a rate of about 40 kilometres (25 miles) a day. Delays arise when the swallows unexpectedly meet cold weather: they may have to mark time or even turn back.

The inactivity of these sanderlings is in contrast to their hurried flight back to the Arctic in spring. Staging posts used in autumn will be bypassed so that breeding can start as soon as possible.

ABOVE *The redshank nests in temperate latitudes and returns to its breeding grounds as much as two months before waders nesting in the Arctic. It then waits one month or more before laying its eggs.*

Enormous numbers of birds pass through some staging posts on the spring migration: 20 million waders visit the Copper River Delta system in Alaska. Among the coasts and islands of the Arctic, large pools of open water, called polynyas, are important for concentrations of sea ducks and auks. Polynyas form where water currents flowing around headlands or islands, and over shallows, hinder the freezing of the sea. They remain unfrozen throughout winter or open up in early spring to create 'oases' where migrating birds can roost safely and dive for food until the waters around their nesting cliffs and coastal tundra are clear of ice.

—— A BOTTLENECK FOR CRANES ——

The Platte River in Nebraska is the meeting place for millions of ducks and geese and forms a bottleneck for 80 per cent of all sandhill cranes on their journey north. The sandbars and shallows make a safe roost for the cranes, which feed in the wet meadows of the surrounding prairies and, increasingly, on cultivated fields. The meadows provide protein-rich insect food, while the fields are a new source of energy-rich spilt grain. The first cranes arrive at the Platte River in early February and the last depart two months later. The wait is necessary because the nesting grounds further north are still in the grip of winter, and the cranes delay their arrival until the last of the snow is melting and the ground is soft enough for them to probe for insects.

RETURNING SEQUENCE

It is a general rule that the males of migratory species return to the breeding ground before the females. Their need is to establish a territory and be in position, singing and displaying, when the females arrive. (The argument is similar to that for males of partial migrants remaining on the breeding grounds through the winter: see page 31.) As there is competition for the best territories, with some males perhaps failing to gain any territory and forfeiting the right to breed, there is a race northward in spring. To this end, the males of some species have evolved a wing shape that is efficient for fast flight (see page 93). Male willow warblers, for example, have longer, more pointed wings than females. The priority for females is that they should be more efficient at gathering food, and their rounded wings give them greater manoeuvrability as they forage among foliage. Nevertheless, females should not lag too far behind the males because of competition to mate with the best males.

Among species in which nesting does not commence until the bird is several years old, youngsters do not return to the breeding grounds until they are mature. Some stay in the winter home through their first summer and move only part of the way to the nesting areas in subsequent years. For

A flock of male red-winged blackbirds on the move. These North American migrants are racing back northwards to establish territories two or three weeks before the less colourful females arrive.

instance, one-year-old ospreys remain in their winter home and two-year-olds spend summer further north; the birds do not reach the breeding grounds until they are older. For species that normally nest in their first spring, young birds are likely to return later than older birds, perhaps because their inexperience has resulted in slower preparation for the journey and less well-orientated flight. A difference of one week in arrival time between young and old male American redstarts can be seen easily because of their different plumage. Fully adult males are black and orange but younger males are brown and yellow like the females. These young males sometimes manage to breed but they usually acquire second-rate territories and their nesting success is low.

One of the advantages of an early return is that a bird can reclaim its old territory. Familiarity with a place from the previous year saves time in searching for food and nesting

RIGHT *If food is scarce, the snowy owl may have to move its territory and search for a new home with an abundance of prey.*

BELOW *This female American redstart will probably have mated with an older male which returned before the younger males.*

places, and gives birds confidence in fending off other settlers. Fidelity is seen at its extreme in some large, long-lived birds-of-prey whose nests are enlarged with new material each spring and develop into such obvious features of the countryside that they become landmarks. Creag Iolaire (Gaelic for Eagle Rock) on maps of the Highlands of Scotland records the location of many traditional nesting places of the golden eagle. For much the same reason of familiarity, young birds of many species return to nest in the neighbourhood where they were raised. Young peregrines have been seen visiting the very ledge where they were reared – and being chased away by their parents. From their early experiences and wanderings (see page 43), they will have acquired some knowledge which will help them in their choice of nesting places.

Exceptions to this trend are seen among those finches that feed on seeds when nesting. As the locations of good seed crops vary, so the birds shift their nest sites. This is particu-larly seen in species that irrupt (see page 37), and established nesters as well as novices are forced to wander. Some finches routinely move their nest sites between successive broods in one season. For instance, redpolls normally nest in the birches that form the northern fringes of Scandinavian forests. They may stop to nest in the forests to raise a family if there is a good crop of cones, and then proceed to the birch zone to lay a second clutch of eggs.

There is also the phenomenon of *abmigration*. It is a feature of many ducks, in which ringing has shown individuals nesting

RIGHT *Unlike many species, the redpoll is a nomadic breeder that moves its nesting place to a site where there is a good food supply.*

BELOW *The bullfinch's diet of buds allows it to spend the winter near its breeding ground. Only in the north of Europe is it forced to migrate.*

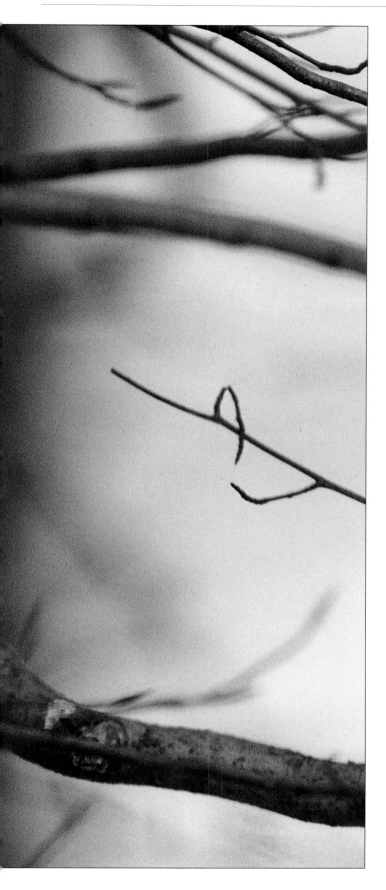

in places far from where they were raised. The term was first used to describe the spring migration of British-reared mallards, which are normally sedentary, to nest in northern Europe. Similarly, in the United States, male ducks that have wintered in the south are sometimes found on unfamiliar breeding grounds. In contrast to most other birds, ducks pair in winter and, instead of moving to their natal breeding ground, these 'wayward' individuals follow their mates to their own native homes. Abmigration may also explain the recovery of a turtle dove ringed as a nestling in England and shot in the Ukraine next summer. It had presumably paired up in Africa during the winter and gone to a new home.

COMPLETING THE CYCLE

Once a bird has arrived safely back on its breeding ground the cycle of migration is complete. The preliminaries to breeding may commence within a day, or even start at staging points on the way, so serving to strengthen the concept of migration as an integral part of a bird's life, rather than a habit to be viewed in isolation.

The 'purpose' of a bird, as of any living organism, is to breed and leave offspring carrying its genes into future generations. Migration is a mechanism for improving a bird's individual chances of survival outside the breeding season, and increasing its reproductive success inside it. It is, in evolutionary terms, an alternative to living in a permanent home.

The enormous advantage that flight gives birds is the ability to migrate efficiently over long distances and cross great barriers, such as oceans, mountains and deserts. It is likely that migration has evolved many times among different groups of birds, which would help to explain the great variety of migratory behaviour now being discovered. The annual movement of the Earth on its axis causes the seasons, resulting in changing food supplies. Migration has given birds the opportunity to exploit this fundamental property of life on Earth.

The brambling adjusts its migration to the weather. In cold springs, it shortens its northward flight and nests in pine forests instead of continuing to the birch zone.

LIST OF SCIENTIFIC NAMES

A list follows of scientific names of birds mentioned in the text. North American common names are given in brackets. The scientific names and the sequence of orders and families are compiled in the order given in *A Complete Checklist of the Birds of the World* by Richard Howard and Alick Moore, Academic Press, 2nd edn., 1991.

Order Sphenisciformes

Spheniscidae	emperor penguin	*Aptenodytes forsteri*
	Adélie penguin	*Pygoscelis adeliae*

Order Gaviiformes

Gaviidae	black-throated diver (North America: Arctic loon)	*Gavia arctica*

Order Procellariiformes

Diomedeidae	royal albatross	*Diomedea epomophora*
	black-browed albatross	*D. melanophris*
Procellariidae	great shearwater (North America: greater shearwater)	*Puffinus gravis*
	sooty shearwater	*P. griseus*
	short-tailed shearwater	*P. tenuirostris*
	Manx shearwater	*P. puffinus*
	Audubon's shearwater	*P. lherminieri*

Order Pelecaniformes

Sulidae	white pelican	*Pelecanus onocrotalus*
	gannet (North America: northern gannet)	*Sula bassana*

Order Ciconiiformes

Ciconiidae	white stork	*Ciconia ciconia*

Order Anseriformes

Anatidae	whooper swan	*Cygnus cygnus*
	Bewick's swan	*C. columbianus bewickii*
	pink-footed goose	*Anser brachyrhynchus*
	greylag goose	*A. anser*
	bar-headed goose	*A. indicus*
	snow goose	*A. caerulescens*
	Canada goose	*Branta canadensis*
	barnacle goose	*B. leucopsis*
	brent goose (North America: brant)	*B. bernicla*
	upland goose	*Chloephaga picta*
	shelduck	*Tadorna tadorna*
	green-winged teal	*Anas crecca carolinensis*
	mallard	*A. platyrhynchos*
	eider (North America: common eider)	*Somateria mollissima*

Order Falconiformes

Pandionidae	osprey	*Pandion haliaetus*
Accipitridae	honey buzzard	*Pernis apivorus*
	bald eagle	*Haliaeetus leucocephalus*
	sparrowhawk	*Accipiter nisus*
	goshawk (North America: northern goshawk)	*A. gentilis*
	broad-winged hawk	*Buteo platypterus*
	Swainson's hawk	*B. swainsoni*
	red-backed hawk	*B. polyosoma*
	buzzard	*B. buteo*
	rough-legged buzzard (North America: rough-legged hawk)	*B. lagopus*
	golden eagle	*Aquila chrysaetos*

Falconidae	kestrel	*Falco tinnunculus*
	merlin	*F. columbarius*
	hobby	*F. subbuteo*
	Eleonora's falcon	*F. eleonorae*
	sooty falcon	*F. concolor*
	prairie falcon	*F. mexicanus*
	gyrfalcon	*F. rusticolus*
	peregrine (North America: peregrine falcon)	*F. peregrinus*

Order Galliformes

Tetraonidae	blue grouse	*Dendragapus obscurus*
	ptarmigan (North America: rock ptarmigan)	*Lagopus mutus*
Phasianidae	quail	*Coturnix coturnix*

Order Gruiformes

Gruidae	common crane	*Grus grus*
	sandhill crane	*G. canadensis*
Rallidae	corncrake	*Crex crex*
	Baillon's crake	*Porzana pusilla*

Order Charadriiformes

Haematopodidae	oystercatcher	*Haematopus ostralegus*
Charadriidae	lapwing	*Vanellus vanellus*
	grey plover (North America: black-bellied plover)	*Pluvialis squatarola*
	ringed plover	*Charadrius hiaticula*
	double-banded plover	*C. bicinctus*
	dotterel	*Eudromias morinellus*
Scolopacidae	Hudsonian godwit	*Limosa haemastica*
	bar-tailed godwit	*L. lapponica*
	Eskimo curlew	*Numenius borealis*
	whimbrel	*N. phaeopus*
	spotted redshank	*Tringa erythropus*
	redshank	*T. totanus*
	turnstone (North America: ruddy turnstone)	*Arenaria interpres*
	red-necked phalarope	*Phalaropus lobatus*
	knot (North America: red knot)	*Calidris canutus*
	sanderling	*C. alba*
	dunlin	*C. alpina*
	ruff	*Philomachus pugnax*
Laridae	herring gull	*Larus argentatus*
	lesser black-backed gull	*L. fuscus*
	great black-backed gull	*L. marinus*
	Franklin's gull	*L. pipixcan*
Sternidae	Arctic tern	*Sterna paradisaea*

Order Columbiformes

Columbidae	rock dove/feral pigeon/homing pigeon	*Columba livia*
	woodpigeon	*C. palumbus*
	turtle dove	*Streptopelia turtur*
	mourning dove	*Zenaida macroura*

Order Cuculiformes

Cuculidae	European cuckoo	*Cuculus canorus*

Order Strigiformes

Strigidae	great horned owl	*Bubo virginianus*
	snowy owl	*Nyctea scandiaca*

	long-eared owl	*Asio otus*
	short-eared owl	*A. flammeus*

Order Caprimulgiformes

Caprimulgidae	common nighthawk	*Chordeiles minor*
	nightjar	*Caprimulgus europaeus*

Order Apodiformes

Apodidae	chimney swift	*Chaetura pelagica*
	swift	*Apus apus*
Trochilidae	ruby-throated hummingbird	*Archilochus colubris*
	rufous hummingbird	*Selasphorus rufus*

Order Piciformes

Picidae	wryneck	*Jynx torquilla*

Order Passeriformes

Tyrannidae	eastern phoebe	*Sayornis phoebe*
Hirundinidae	sand martin (North America: bank swallow)	*Riparia riparia*
	swallow (North America: barn swallow	*Hirundo rustica*
	house martin	*Delichon urbica*
Motacillidae	yellow wagtail	*Motacilla flava*
	white wagtail	*M. alba*
Laniidae	red-backed shrike	*Lanius collurio*
Bombycillidae	waxwing (North America: Bohemian waxwing)	*Bombycilla garrulus*
	cedar waxwing	*B. cedrorum*
Troglodytidae	wren (North America: winter wren)	*Troglodytes troglodytes*
Mimidae	northern mockingbird	*Mimus polyglottos*
Turdidae	European robin	*Erithacus rubecula*
	nightingale	*Luscinia megarhynchos*
	bluethroat	*L. svecica*
	redstart	*Phoenicurus phoenicurus*
	whinchat	*Saxicola rubetra*
	stonechat	*S. torquata*
	wheatear (North America: northern wheatear)	*Oenanthe oenanthe*
	grey-cheeked thrush	*Catharus minimus*
	blackbird	*Turdus merula*
	fieldfare	*T. pilaris*
	redwing	*T. iliacus*
	song thrush	*T. philomelos*
	mistle thrush	*T. viscivorus*
Paradoxornithidae	bearded tit (bearded reedling, bearded parrotbill)	*Panurus biarmicus*
Sylviidae	Cetti's warbler	*Cettia cetti*
	sedge warbler	*Acrocephalus schoenobaenus*
	reed warbler	*A. scirpaceus*
	marsh warbler	*A. palustris*
	willow warbler	*Phylloscopus trochilus*
	chiffchaff	*P. collybita*
	wood warbler	*P. sibilatrix*
	Pallas's leaf warbler	*P. proregulus*
	blackcap	*Sylvia atricapilla*
	garden warbler	*S. borin*
	whitethroat	*S. communis*
	lesser whitethroat	*S. curruca*
	barred warbler	*S. nisoria*
	Dartford warbler	*S. undata*
	goldcrest	*Regulus regulus*
Muscicapidae	pied flycatcher	*Ficedula hypoleuca*
Aegithalidae	long-tailed tit	*Aegithalos caudatus*

Paridae	black-capped chickadee	*Parus atricapillus*
	great tit	*P. major*
Certhiidae	treecreeper	*Certhia familiaris*
Emberizidae	Lapland bunting (North America: Lapland longspur)	*Calcarius lapponicus*
	snow bunting	*Plectrophenax nivalis*
	fox sparrow	*Passerella iliaca*
	white-crowned sparrow	*Zonotrichia leucophrys*
	white-throated sparrow	*Z. albicollis*
	dark-eyed junco	*Junco hyemalis*
	savannah sparrow	*Passerculus sandwichensis*
	Ipswich sparrow	*P. s. princeps*
	dickcissel	*Spiza americana*
	indigo bunting	*Passerina cyanea*
Parulidae	golden-winged warbler	*Vermivora chrysoptera*
	tropical parula	*Parula pitiayumi*
	yellow warbler	*Dendroica petechia*
	chestnut-sided warbler	*D. pensylvanica*
	Cape May warbler	*D. tigrina*
	Blackburnian warbler	*D. fusca*
	magnolia warbler	*D. magnolia*
	yellow-rumped warbler	*D. coronata*
	blackpoll warbler	*D. striata*
	bay-breasted warbler	*D. castanea*
	American redstart	*Setophaga ruticilla*
	northern waterthrush	*Seiurus noveboracensis*
	hooded warbler	*Wilsonia citrina*
	yellow-crowned redstart	*Myioborus flavivertex*
	russet-crowned warbler	*Basileuterus coronatus*
Icteridae	red-winged blackbird	*Agelaius phoeniceus*
	bobolink	*Dolichonyx oryzivorus*
Fringillidae	chaffinch	*Fringilla coelebs*
	brambling	*F. montifringilla*
	serin	*Serinus serinus*
	siskin	*Carduelis spinus*
	pine siskin	*C. pinus*
	redpoll (North America: common redpoll)	*Acanthis flammea*
	house finch	*Carpodacus mexicanus*
	common crossbill (North America: red crossbill)	*Loxia curvirostra*
	bullfinch	*Pyrrhula pyrrhula*
	evening grosbeak	*Coccothraustes vespertinus*
Ploceidae	red-billed quelea	*Quelea quelea*
Passeridae	house sparrow	*Passer domesticus*
Sturnidae	starling (North America: European starling)	*Sturnus vulgaris*
Corvidae	jay	*Garrulus glandarius*
	nutcracker	*Nucifraga caryocatactes*
	hooded crow	*Corvus corone cornix*

FURTHER READING

For a non-technical introduction to bird migration and related topics I recommend the following books:

Alerstam, Thomas. *Bird Migration*, Cambridge University Press, 1990

Baker, R. Robin. *The Evolutionary Ecology of Animal Migration*, Hodder & Stoughton, 1978

Baker, R. Robin (Ed.) *The Mystery of Migration*, Macdonald, 1980

Baker, R. Robin. *Bird Navigation: the Solution of a Mystery?* Hodder & Stoughton, 1984

Burton, Robert. *Bird Flight*, Facts on File, 1990

Dorst, Jean. *The Migrations of Birds*, Heinemann, 1962

Durman, Roger, (Ed.), *Bird Observatories in Britain and Ireland*, T & A. D. Poyser, 1976

Elkins, Norman. *Weather and Bird Behaviour*, T & A. D. Poyser 2nd edn., 1988

Hale, W. G. *Waders*, (New Naturalist series, 65), Collins, 1980

Mead, Chris, *Bird Migration*, Country Life Books, 1983

Morse, Douglass H. *American Warblers: an Ecological and Behavioral Perspective*, Harvard University Press, 1989

Newton, Ian. *Finches* (New Naturalist series, 55), Collins, 1972

Ogilvie, M. A. *Wild Geese*, T & A. D. Poyser, 1978

Pearson, Bruce. *An Artist on Migration*, Harper-Collins, 1991

Terborgh, John. *Where Have All the Birds Gone? Essays on the Biology and Conservation of Birds That Migrate to the American Tropics*, Princeton University Press, 1989

Walter, H. *Eleonora's Falcon: Adaptations to Prey and Habitat in a Social Raptor*, University of Chicago Press, 1979

For those wanting to take their reading further, the following books provide more detailed and technical information on various aspects of bird migration:

Gwinner, E. (Ed.), *Bird Migration: Physiology and Ecophysiology*, Springer-Verlag, 1990

Keast, A. and Morton, E. S. (Eds.), *Migrant Birds in the Neotropics*, Smithsonian Institution Press, 1980

Kerlinger, Paul. *Flight Strategies of Migrating Hawks*, University of Chicago Press, 1989

Moreau, R. E. *The Palaearctic-African Bird Migration Systems*, Academic Press, 1972

The following list of papers published in scientific journals is not comprehensive but gives an introduction to the literature on bird migration:

Adriaensen, F. and Dhondt, A. A. 1990. Population dynamics and partial migration of the European robin (*Erithacus rubecula*) in different habitats. *Journal of Animal Ecology* 59: 1077–1090

Alerstam, T. 1975. Crane *Grus grus* migration over sea and land. *Ibis* 117: 489–495

Axell, H. E. et al. 1963. Finch migration at Minsmere. *Bird Notes* 30: 181–186

Bibby, C. J. and Green, R. E. 1981. Autumn migration strategies of Reed and Sedge Warblers. *Ornis Scandinavica* 12: 1–12

Biebach, H., Friedrich, W. and Heine, G. 1986. Interaction of bodymass, fat, foraging and stopover period in trans-Sahara migrating passerine birds. *Oecologia* 69: 370–379

Biebach, H. 1983. Genetic control of partial migration in robins. *Auk* 100: 601–606

Bourne, W. R. P. 1980. The midnight descent, dawn ascent and reorientation of land birds migrating across the North Sea in autumn. *Ibis* 122: 536–540

Evans, P. R. 1966. An approach to the analysis of visible migration and a comparison with radar observations. *Ardea* 54: 14–44

Gudmundsson, G. A., Lindström, Å. and Alerstam, T. 1991. Optimal fat loads and long-distance flights by migrating Knots *Calidris canutus*, Sanderlings *C. alba* and Turnstones *Arenaria interpres*. *Ibis* 133: 140–152

Hagan, J. M., Lloyd-Evans, T. L. and Atwood, J. L. 1991. The relationship between latitude and the timing of spring migration of North American landbirds. *Ornis Scandinavica* 22: 129–136

Hedenström, A. and Pettersson, J. 1986. Differences in fat deposits and wing-pointedness between male and female willow warblers on spring migration at Ottenby, SE Sweden. *Ornis Scandinavica* 17: 182–185

Helbig, A. J. 1991. Inheritance of migratory direction in a bird species: a cross-breeding experiment with SE- and SW- migrating blackcaps (*Sylvia atricapilla*). *Behavioral Ecology and Sociobiology* 28: 9–12

Lavee, D., Safriel, U. N. and Meiljson, I. 1991. For how long do trans-Saharan migrants stop over at an oasis? *Ornis Scandinavica* 22: 33–44.

Lindström, Å. 1991. Maximum fuel deposition rates in migrating birds. *Ornis Scandinavica* 22: 12–19

Lovei, G. L. 1989. Small birds have enough fat to cross Sahara in one hop. *Current Ornithology* 6: 132–174

Massman, D. and Klaasen, M. 1978. Changes in fuel reserves on a journey. *Auk* 104: 603–616

Moore, F. R. 1977. Geomagnetic disturbance and the orientation of nocturnally migrating birds. *Science* 196: 682–684

Pearson, D. J. and Backhurst, G. C. 1976. The southward migration of Palaearctic birds over Ngulia, Kenya. *Ibis* 118: 78–105

Pennycuick, C. J., Alerstam, T. and Larsson, B. 1979. Soaring migration of the Common Crane *Grus grus* observed by radar and from an aircraft. *Ornis Scandinavica* 10: 241–251

Perdeck, A. C. 1974. An experiment on the orientation of juvenile Starlings during spring migration. *Ardea* 62: 190–195

Piersma, T. and Jukema, J. 1990. Energy budget of a godwit between Africa and Holland. *Ardea* 78: 315–337

Piersma, T., Zwarts, L. and Bruggemann, J. H. 1990. Behavioural aspects of the departure of waders before long-distance flights: flocking, vocalizations, flight paths and diurnal timing. *Ardea* 78: 78–90

Porter, R. and Willis, I. 1968. The autumn migration of soaring birds at the Bosphorus. *Ibis* 110: 520–536

Rappole, J. H. and Warner, D. W. 1976. Relationships between behaviour, physiology and weather in avian transients at a migration stopover site. *Oecologia* 26: 193–212

Richardson, W. J. 1978. Timing and amount of bird migration in relation to weather: a review. *Oikos* 30: 224–272

Ulfstrand, S. 1963. Ecological aspects of irruptive bird migration in northwestern Europe. *Proceedings of the 13th International Ornithological Congress* 780–794

Wege, M. L. and Raveling, D. G. 1984. Flight speed and directional responses to wind by migrating Canada geese. *Auk* 101: 342–348

INDEX

Page numbers in *italics* indicate captions to illustrations

A

abmigration 150–3
Africa, winter home 121, 124, 128–32
aircraft, tracking by 16, 96–7
albatross 91, 95
  black-browed 111, *111*
  royal *91*
Alerstam, Thomas 16, 66, 96
altitude 10, 19, 83, 89
  reached on thermals 96
America, winter homes 132–5
Antarctic 83, *83*
Arctic regions 20, 23, 83, 142–3
Aristotle 13, 20, 33
Audubon, John James 15
auks 46, 48, 146

B

Balsfjord, Norway *7*
Baltic Sea 145
Barents, Willem 13
barnacles 13
Bartram, William 53
bats 19
bee-eaters 57, 91
Benguela Current 83
Bertolet, Col de 14
bird-banding *see* ringing
bird-ringing *see* ringing
bird-watching 13–14
birds-of-prey *see* predators
blackbird 120
  red-winged *147*
blackcap 30, 68–9, *69*, 93
bluethroat 99, 108
bobolink 133, *133*
Bosporus 33, 79
brambling 113, *153*
brant *see* goose, brent
breathing system 19
breeding 42, 153
bullfinch *150*
bunting 57, 108
  indigo 60, 65, 70–2, *72*
  Lapland 105
  snow 46, 138, *138*
buzzard 27
  honey 114
  rough-legged 37

C

Cape May, Connecticut 14, 27, 79
Cape Morris Jesup 83
Cape Wrath 53, 142
captive birds, migratory
  restlessness 60, 68–9, 125
carbohydrates 18
chaffinch 29, *29*, 31, 66, 69, 87, 113
chickadee, black-capped *23*
chiffchaff 68, 93
classification system 31
clock, internal 49, 64, 140
clouds
  birds not flying in 51
  flying above 89
  navigation in 64–5, 72

Common Birds Census *121*
compass course 63–6
  magnetic 65–6, *66*, 70, 72
  by moon 65, 70
  by stars 63, 65, 70–2
  by sun 60, 63–5, *65*, 70, 72
conservation 28, 114–15, 126
corncrake 11
courtship *138*
crab, horseshoe 101
crake, Baillon's *107*
crane 13, 16, 46, 54, 57, 72, 93, 95
  common 87, 96–7, *97*, 125, *138*
  sandhill *63*, 109, 146
crèches 48
crossbill 37, 118–20
  common *123*
crow 26
  hooded 67
cuckoo 55, 60, *60*
curlew, Eskimo 34, 44, 114

D

Dardanelles 95
daylength, effect on migration 140
death rate 33, 73
Delaware Bay 33, 79, 101, *101*, 145
deserts 79, 80, *80*
destinations 13–14
dickcissel 23
distances travelled 18–19, 26–8
  per day 98
diurnal (daytime) migration 55, 57
diver 48
  black-throated 34
Doppler effect 63
dotterel 110, *110*
doughbird *see* curlew, Eskimo
dove 114
  mourning 104
  rock 60
  turtle 13, 130, *130*, 153
drag 19, 89, 91, 95
droughts 76, 105, 108
Duck Stamps 99
ducks 46, 48, 91, 121, 125, 146, 150–3
  sea 146
Duncan, Martin 108
dunlin 16, *41*, 46, 98, 126, *126*
dyes, to identify birds 16

E

eagle
  bald *89*
  golden 150
eider 87
Ellesmere Island 16
Emlen cage *72*
energy consumption
  in relation to height 89
  in relation to speed 84
  from muscle 19
  used in flight 18–19
environment, effect on migration 49
estuaries 115, 126
EURING (European Union for Bird
  Ringing) 15
eyes 64

F

Fair Isle 14
falcon 113, 114
  Eleonora's 113
  prairie 11
  sooty 113
falls 13, 106, 110
Falsterbo 14, 33, 79
fat
  breeding success in relation to 143
  change of diet for 44
  for energy 18–19
  metabolizing 19
  storing *106*
feathers 46
females
  migrating alone 31, 33
  wing shapes 147
fieldfare 39, 73, 120
finches 26, 37, 46, 51, 57, 95, 150
  house 31
flamingo 48
flight
  bounding 95
  duration 18–19
  dynamics 87, *87*
  energy used 18
  factors aiding 90–7
  in formation 51, 83, 95, *95*, 137
  height 10, 19, 80, 89
    reached on thermals 96
  labour-saving 90–7, 145
  non-stop 17, 80
  plans 75–101
  power in relation to weight 19
  power-speed curve *87*
flight speed 17, 80, 84–8, 96–7, 145
  average 98
  maximum range 84, 87–8
  minimum power 84, 88
  in relation to fuel
    consumption 84
  style for migration 83
  undulating 93, 95
flightlessness, during moult 46, 48
flycatcher 20, 55, 115
  pied 129
fog
  flying above 89
  navigation in 72
food
  during journey 99–100
  migration for 20–3, 37–8
  in preparation for journey 44, 140
Forbes-Watson, Alec 130
forests 132
  destruction 115, 123
  rain 129, 132
formation flying 51, 83, 95, *95*, 137
fuel
  consumption 12
    in relation to speed 84, 87, *87*
  for energy 18–19
  oxidizing 19
  in preparation for journey 44
  relationship to body weight 19
  replenishing on journey 99–100
Fundy, Bay of 145

G

gamebirds 11
gannet 111
Gätke, Heinrich 14
geese 19, 46, 48, 79, 93, 95, 98, 142–3, 146
  bar-headed 79
  barnacle 13, *13*, 46, *54*, 60, 113, *143*
  brent 142–3
  Canada 17, 36, 72, 95, *142*
  greylag 53–4
  pink-footed 10, 48, *51*, 53–4
  snow *84*, *137*, *142*
  upland 7
Gibraltar 33
Gibraltar, Strait of 79, 95
gliding 93, 95
glycogen 18
gnomic projection 77, *78*
godwit
  bar-tailed 10, 26, 90
  Hudsonian 34
goldcrest 110
goshawk 37
Great Circle route 66, 69, 77–8, *77*, *78*, 79, 111
grebes 46
Greenland 77, 78, 79, 138
grosbeak, evening 38
grouse 37
  blue 11
guillemot 46
Gulf Stream 39
gull 121
  Franklin's 59
  great black-backed 113
  herring 113
  lesser black-backed *43*
gyrfalcon 16, 113

H

harriers 124
Harrison, William 13
hawk 27, 37, 72, 93, 95, 118
  red-backed 79
  rough-legged *see* buzzard,
    rough-legged
  Swainson's 94–5, *94*, 133
Hawk Mountain, Pennsylvania 14, 27, 33, 93
hazards 103–15
Hebrard, James 51
Heligoland 14, 48
Heligoland trap *15*
herons 57, 93, 95, 114
hibernation 13
hides *14*
Himalayas 79
hobby 129
Hudson, W. H. 7, 10, 42
human threat 114–15
hummingbirds 118, 123
  ruby-throated 124
  rufous 11, 100, *100*, 118

I

ice-caps 78, 79
Iceland 79

infra-sounds, navigation by 63
inheritance, programmed routes 68–9
injured birds, attempting to migrate 7
instinct 7
　navigation by 60
　　programmed routes 68–9
irruptions 37–8
islands 13

J

jay 37
junco, dark-eyed 30, *30*, 31
juveniles
　competition with adults 123
　in crèches 48
　exploration 43, 66
　leaving with females 31
　navigation 60
　navigation errors 111
　remaining in winter homes 147–8

K

kestrel 36, 113
kite 13
kiwi 63
knot 7, 16, *25*, 26, 28, *28*, 101, *101*,
　145
Knuth, Eigil 79
Kramer, Gustav 64, 65
Kramer cage *65*

L

Labrador Current 34
Lack, David 51
Landes, France 96
landmarks, navigation by 43, 63, 70,
　72
lapwing 73, *73*, *106*
Larson, Gary, cartoon 18
light, polarized 64–5
lighthouses 110
Linnaeus (Carl von Linné) 31
longspur, Lapland 105
loon *see* diver
　Arctic *see* diver, black-throated
Lowery, George H. 14

M

magnetic compass 65–6, *66*, 70, 72
magnetite 66
males
　not migrating 31
　returning before females 31, 147
　wing shapes 147
mallard 60, *60*, 64, 65, 153
martin 57
　bank 53
　house 104, 105
　sand 43, *43*, 104, *104*, 105, 118
Matthews, G. V. T. 60
merlin 113, *113*
metabolism 19
Mexico, Gulf of 79
migration
　barriers 79–80
　destinations 13–14
　differential 31–3
　distances 18–19, 26–8
　irruptions 37–8
　journey times 98
　leap-frog 36
　long-distance 26–8, 34, 83
　loop 33–4

meaning 11
partial 29–30
preparations for 41–57, 140
reasons for 20–3
restlessness for 60, 65, 66, 68–9,
　125
return 137–53
reverse 51
routes 27
short journeys 11
strategy for 90–7, 145
timing 49–57, 140
tracking 13–17
types of 25–39
understanding 12
Minsmere 51
mist nets *15*
mockingbird, northern 29
moon, navigation by 65, 70
moon-watching 14
Moreau, Reg 129–30
Moreau's Paradox 129–30
Morecambe Bay 126
Mortensen, Hans Christian 15
moult 33, 42, 45–8
mountain ranges 79
murres *see* guillemot
muscles 19

N

natural selection 73
Naumann, Johann 68
navigation 18, 59–73
　errors of 60, 73, 107, 108, 111
　methods 63–7
　programmed routes 68–9
　reorientation 67
　studying 60
nesting 138, 142–3, 145, 147–8, 150
nets, mist *15*
Newton, Ian 129
Ngorongoro Crater, Tanzania *132*
nighthawk *11*
　common 133
nightingale 129
nightjar 20
nocturnal migration 10, 13, 51, 55–7,
　72
nomads 118–20, 128, *150*
nutcracker 37, *49*

O

observatories 14
oil pollution *114*, 115
oil rigs 108
oriole 129
osprey *88*, 148
owl 11, 37
　great horned 37
　long-eared 76, 113
　short-eared 113
　snowy 37, 148
oystercatcher 16, *48*

P

Papi, Professor Floriano 63
Paris, Matthew 37
parula, tropical 135
pelican 57, 93, *93*, 95
penguin
　Adélie 18, 64, *64*
　Emperor 18
Pennycuik, Colin 16, 96
Penrose, Harald 16

Perdeck, A. C. 66—7
peregrine *17*, 105, 133, 150
Pernau, Baron von 42
Persian Gulf 115
Peterson, Roger Tory 135
petrels 95
phalarope 31
　red-necked *31*
phoebe, eastern 15
pigeon 57
　homing 60, 63, 65, 66, 67, 70
　racing 70
　wood 26
piloting 63, 70, 72
pipits 57, 108
Platte River, Nebraska 146
plover 101
　black-bellied *see* plover, grey
　double-banded 7
　grey 46, 123, *124*
　ringed 36
polarization 64–5
Pole Star 65
pollution *114*, 115
polynyas 146
power, relationship to body weight 19
power-speed curve 87, *87*
predators
　attacking migrants 57, 113
　breeding 113
　migration 57, 113
　nesting 150
　size of male and female 31
　in winter 31, 124, 128–9
Project FeederWatch 37
protein 18, 19
ptarmigan 20

Q

quail 106, 114, 130
quelea, red-billed 128

R

radar 14, 50–1, 76, 80, 89, 113
radio tagging 16–17, 76
rails 46
rainfall, effect on migration 130
record-breakers 19
redpoll 150, *150*
redshank *146*
　spotted *103*
redstart 13, 33, 110
　American 93, *124*, 148, *148*
　yellow-crowned 93
redwing 120, *120*
restlessness, migratory 60, 65, 66,
　68–9, 125
rhumbline 77–8
ringing 14–16, *16*, 76, 128, 130
　rings used 15–16, *16*
　trapping birds for *15*
robin 13, 26, 65
　European 32–3, *32*, 70, 72
roosts 43, 99–100
routes
　barriers on 79–80
　Great Circle 66, 69, 77–8, *77*, *78*,
　　79, 111
　programmed 68–9
ruff 78, *78*

S

Sahara 34, 57, 80, *80*
Sahel 115, 129–30, 140

sanderling 145, *145*
sandpiper 101
sandstorms 89
Scilly Isles 108, *108*
sea
　birds drowning in 76, 105, 110
　narrow crossing points 79
　navigating by waves 72, 97
seabirds *12*, 33–4, 95
serin 20–3
Shasta, Mount, California *84*
shearwater, 91, 95
　Audubon's 87
　great 33
　Manx 33
　short-tailed *33*
　sooty *34*
shelduck 48, *48*
ships, birds travelling on 105, 108
shrike 113
　red-backed 34
siskin 38
　pine 37
smell, navigation by 63
snowbird *see* junco, dark-eyed
soaring 93–5
songbirds .
　flying height 89
　human threat to 114, 115
　journey time 98
　mailing 46
　navigation 60
　return journey 140
　routes 79
　wings 93
　in winter 129
　see also individual species
sound, navigation by 63
sparrow
　American 55, 108
　fox 29, 36, *36*
　house 11
　savannah 72
　white-crowned *70*
　white-throated 64, 70, 88
sparrowhawk 113
speed 17, 80, 84–8, 96–7, 145
　adjusted to wind 88
　average 98
　fuel consumption in relation to 84
　maximum range 84, 87–8
　minimum power 84, 88
staging posts 99–101, 115, *115*, 145,
　146
starling 15, 29, *57*, 60, 64, *65*, 66–7,
　67
stars, navigation by 63, 65, 70–2
stonechat 23
stork 57, 93
　white *55*, 57, 96, 104, *117*, 118,
　　132
sun, navigation by 60, 63–5, *65*, 70,
　72
Svalbard 13, 20
swallow 15, 16, 20, 23, 104, 105,
　118, 138, 143
　hibernation theory 13
　preparing for migration 42, *42*, 51,
　　57
　return journey 145
　species
　　bank *see* martin, sand
　　barn 44, 118, 133
　　cliff 133

house 53
wings 91
in winter 120, 128, 129
swan 19, 46, 93, 95, 98
Bewick's *18*
whooper 19, *27*
Swath, H. S. 36
swift 20, 87, 91
chimney 106–7

T

teal, green-winged *125*
temperatures, extremes of 20, *26*
tern 91, 104
Arctic 33–4, *75*, 83, *83*
territories
competition for 31, 33, 100, 147–8
moving 150
reclaiming 123, 148–50
in winter homes 123, 128
thermals 55, 57, 89, 93–5, 96–7
thrush 55, 113, 118, 120
grey-cheeked 108, *109*
mistle 120
song *114*, 120
Tinbergen, Niko 138
tits 20, 37
bearded 20
great *20*
long-tailed 29
tracking 13–17, 96–7
transmutation 13
trapping, for ringing *15*
treecreepers 20
tropics, wintering in 128–35, 140

tundra 23, 28, 48
turnstone *9*, *46*, *98*, 101, 145

U

ultra-violet light 64
US Fish and Wildlife Service 15

V

vagrants 107, 108, 109, 111, *111*
vireos 115

W

Wadden Sea 48, 126, 145
waders 19, 31, 36, 46, 50, 120, 121
flight 98
flying height 89
predators 113
preparating for departure 42, 51, 54
routes 34, 79, *99*
staging posts 99–100, 101, 113, 115, 145, 146
wings 91
in winter 26, 31, 39, 125, 126, 133, 140
see also individual species
wagtail 80
white 128
yellow 80, 128
warbler 10, 20, 42, 55, 65, 80, 98, 104, 108, 113, 118
barred 64
bay-breasted 123
Blackburnian 93
blackpoll 34, 111, 140

Cetti's *91*, 93
chestnut-sided 123, 128
Dartford 93
garden 60, *68*, 70, 72, 128, *128*
golden-winged *53*
hooded 135
magnolia 128
marsh 128
Pallas's leaf 111
reed 44, *44*
russet-crowned 93
sedge 19, 44, 93, 132
willow 60, 68, 93, *129*, 147
wood
American 100, 115, 128, 133, 135, *135*, 140
European *132*, 135, 145
yellow 93, *118*, 128
yellow-rumped 125, 135, *135*, 140
warfare 115
Wash 126
waterbirds 39
waterthrush, northern 100, 124
waxwing *38*
cedar 38
weather
conditions preferred 54
effect on survival 104–5
migration affected by 53–4, *90*
migration caused by 39
weaverbird 128
weight
breeding success in relation to 143
relationship of fat to 19
West Indies 121

wetlands 115, 123
wheatear *26*, 42, *77*, 78, 80, *80*, 108, 114
whimbrel 140, *140*
whinchat 23, *23*, 110
White, Gilbert 31
Whitefish Point 79
whitethroat 72, 121, *121*
lesser 46
wildfowl, refuges 99
wind
adjusting speed to 88
compensation for 72
dangers of 106–7
drifting with 72
gusts 89
head 88
migration assisted by 53–4, 90, *90*
migration routes affected by 33–4
naming 33
prevailing *33*, 82
side, adjustment for 17, 72
systems 33
tail *26*, 82, 88, 90
wings
designs 90–3
male/female differences 147
winter homes 117–35
woodpecker 11, 20, 105
woodpigeon 26
wren 11, 26, 55, 105
wryneck 110

Z

Zugunruhe 60, 65, 66, 68–9, 125

ACKNOWLEDGEMENTS

AUTHOR'S ACKNOWLEDGEMENTS
Information for this book has been gathered from a number of books and many scientific journals. The list of key publications shows my indebtedness to the authors for their fascinating accounts of many aspects of bird migration which I have recorded in these pages. I am also grateful to Nick Davidson and Bill Hankinson for advice.

ARTWORK
Eddison Sadd and the author have taken all reasonable steps to obtain permission to use sources of artwork. We thank the authors and publishers of the following books and journals: 28 Piersma, T. and Davidson, N. C. *Wader Study Group Bulletin 63, Supplement*, 1991; 37 Dunn, E. H. *FeederWatch News*, vol 3, Cornell Laboratory of Ornithology, 1990; 43 Mead, C. J. and Harrison, J. D. *Bird Study*, vol 26, Blackwell Scientific Publications, 1979; 66 Alerstam, Thomas *Bird Migration*, Cambridge University Press, 1990; 67 Perdeck, A. C. *Ardea*, vol 46, Netherlands Ornithologists' Union, 1958; 68 Gwinner and Wiltschko, W. *Journal of Comparative Physiology*, vol 127, Springer-Verlag, 1978; 69 Helbig, A. J. *Behavioural Ecology and Sociobiology*, vol 28, Springer-Verlag, 1991; 78 Alerstam, Thomas *Bird Migration*, Cambridge University Press, 1990; 99 Morrison, R. I. G. and Myers, J. P., Canadian Wildlife Service 'Shorebird Flyways in the New World', in *Flyways and Reserve Networks for Waterbirds*, eds. Boyd, H. and Pirot, J. Y., International Waterfowl and Wetlands Research Bureau (IWRB) Special Publications, No 9, 1989, used with permission of the Minister of Supply and Services Canada, 1992; 109 Elkins, Norman *British Birds*, vol 72, Macmillan, 1979; 121 Marchant, John H., Hudson, Robert, Carter, Steve P. and Whittington, Phil *Population Trends in British Breeding Birds*, British Trust for Ornithology, 1990; 135 Keast, Allen and Morton, Eugene S. *Migrant Birds in the Neotropics: Ecology, Behavior, Distribution and Conservation*, Smithsonian Institution Press, 1980.

PHOTOGRAPHERS

t top   b below   l left

2 Roger Tidman / FLPA; 6–7 Nick Davidson; 8–9 Roger Tidman / NHPA; 10 Richard Mearns / Nature Photographers; 11 Chris Knights; 12 Andrew Cleave / Nature Photographers; 13 Chris Knights; 14 Roger Tidman / FLPA; 16 D. C. Twichell / Manomet Bird Observatory; 17 Eric & David Hosking; 18 Chris Knights; 19 Eric & David Hosking; 20–21 Nature Photographers; 22 S. Krasemann / NHPA; 23 Laurie Campbell / NHPA; 24–25 Roger Tidman / Nature Photographers; 26 Gertrud & Helmut Denzau; 27 Chris Knights; 28 A.N.T. / NHPA; 29 Eric & David Hosking / FLPA; 30 R. Austing / FLPA; 31 Winfried Wisniewski; 32 Roger Tidman / Nature Photographers; 34–35 Eric & David Hosking / FLPA; 36 S. Maslowski / FLPA; 37 S. Maslowski / FLPA; 38 David Tipling; 39 Maurice Walker / FLPA; 40–41 David Tipling; 42 Peggy Heard / FLPA; 43 J. Watkin / FLPA; 44–45 Gertrud & Helmut Denzau; 46 Paul Sterry / Nature Photographers; 47t Roger Wilmshurst /FLPA; 47b Roger Tidman / FLPA; 48 Flip de Nooyer / Bruce Coleman; 49 David Tipling; 50–51 Chris Knights; 52–53 David Tipling; 54 Roger Wilmshurst / FLPA; 55 Manfred Danegger; 56–57 Hellio & Van Ingen / NHPA; 58–59 Michael & Patricia Fogden; 60 Manfred Danegger; 61 Chris Knights; 62 S. Krasemann / NHPA; 64 C. Carvalho / FLPA; 69 Konrad Wothe / Bruce Coleman; 70–71 Chris Knights; 73 Roger Tidman / NHPA; 74–75 Richard Mearns / Nature Photographers; 76 Chris Knights; 77 A. R. Hamblin / FLPA; 78 Nature Photographers; 79 Eric & David Hosking; 80 Eric & David Hosking; 82 Gertrud & Helmut Denzau; 83 Ashod Papazian / NHPA; 84–85 Chris Knights; 86–87; Paul Sterry / Nature Photographers; 88 H. Clark / FLPA; 89 Steve McCutcheon / FLPA; 91t FLPA; 91b Roger Tidman / Nature Photographers; 92–93 Paul Sterry / Nature Photographers; 94 W. S. Clark / FLPA; 95 Geoff du Feu / Nature Photographers; 96–97 Márquez / Bruce Coleman; 98 Laurie Campbell / NHPA; 99 David Tipling; 100 Bob & Clara Calhoun / Bruce Coleman; 101 G. Moon / FLPA; 102–103 Gertrud & Helmut Denzau; 104 Roger Tidman / NHPA; 105 G. Cripps / British Antarctic Survey; 106 Paul Sterry / Nature Photographers; 107 David Tipling; 108 Roger Tidman / Nature Photographers; 109 Eric & David Hosking / FLPA; 110 Roger Tidman / FLPA; 111 S. Maslowski / FLPA; 112 B. S. Turner / FLPA; 114t Andrew Cleave / Nature Photographers; 114b Paul Sterry / Nature Photographers; 115 Eric & David Hosking; 116–117 Winfried Wisniewski; 118–119 Paul Sterry / Nature Photographers; 120 E. A. Janes / NHPA; 121 Kevin Carlson / Nature Photographers; 122–123 David Tipling; 124 Roger Wilmshurst / FLPA; 125 Paul Sterry / Nature Photographers; 126–127 R. Thompson / FLPA; 128 Jan Van de Kam / Bruce Coleman; 129 Konrad Wothe / Bruce Coleman; 130–131 Manfred Danegger; 132t Winfried Wisniewski; 132b Konrad Wothe / Bruce Coleman; 133 Kevin Carlson / Nature Photographers; 134 Wayne Lankinen / Bruce Coleman; 136–137 Chris Knights; 138 Roger Wilmshurst / FLPA; 139 Liz & Tony Bomford; 140–141 W. Wisniewski / FLPA; 142 Norbart Rosing / Bruce Coleman; 143 Chris Knights; 144–145 Eric & David Hosking / FLPA; 146 Paul Sterry / Nature Photographers; 147 FLPA; 148l L.West / FLPA; 148–149 Eero Murtomäki / NHPA; 150 A. Wharton / FLPA; 151 FLPA; 152–153 Chris Knights

EDDISON SADD
Editor Jonathan Elphick
Proof Readers Yvonne Ibazebo and Sam Merrell
Designer Nigel Partridge
Artists Hardlines
Picture Researcher Liz Eddison
Indexer Dorothy Frame
Production Hazel Kirkman and Charles James
Editorial Director Ian Jackson
Creative Director Nick Eddison